The Cambridge Technical Series
General Editor: P. Abbott, B.A.

TECHNICAL HANDBOOK
OF
OILS, FATS AND WAXES

VOLUME I

Chemical and General

TECHNICAL HANDBOOK
OF
OILS, FATS AND WAXES

BY

PERCIVAL J. FRYER, F.I.C., F.C.S.

Soap and Glycerine Works Manager and Chief Chemist
Lecturer in Oils, Fats and Waxes at the Polytechnic, Regent Street, W.
Honours Silver Medallist, City and Guilds of London

AND

FRANK E. WESTON, B.Sc. (1st Hons.), F.C.S.

Head of the Chemistry Department, the Polytechnic, Regent Street, W.
Author of *The Detection of Carbon Compounds* and *Elementary Experimental Chemistry*

WITH 33 ILLUSTRATIONS AND 36 PLATES

VOLUME I

Chemical and General

Cambridge:
at the University Press
1918

CAMBRIDGE
UNIVERSITY PRESS

University Printing House, Cambridge CB2 8BS, United Kingdom

Published in the United States of America by Cambridge University Press, New York

Cambridge University Press is part of the University of Cambridge.

It furthers the University's mission by disseminating knowledge in the pursuit of education, learning and research at the highest international levels of excellence.

www.cambridge.org
Information on this title: www.cambridge.org/9781107687318

First edition 1917
Second edition 1918
First published 1918
First paperback edition 2014

A catalogue record for this publication is available from the British Library

ISBN 978-1-107-68731-8 Paperback

PREFACE

THE present small treatise has been designed primarily, as its title indicates, to meet the need of the technical worker, the works chemist, and others less directly concerned in the technology of the oils, fats and waxes.

It has been the experience of the authors, extending over several years, that, in the case of technical men generally, there exists a wide knowledge of the practical issues of the subject concerned, side by side with much ignorance of the basic principles underlying such issues. In the following pages an endeavour has been made to explain in as simple a manner as possible the theoretical basis upon which the technical processes rest, as well as to describe the various reactions concerning the industry. Such explanations have been mostly printed in specially small type, so that in the case—doubtless of frequent occurrence—of such matter being already familiar, it may be readily passed over by the reader.

For the first time, we believe, a survey has been made of the whole subject of the oils, fats and waxes in a single treatise. An obvious advantage of this is the wider outlook so obtained, and the possibility of comparing within the limits of the one volume, the "natural" oils and fats with the mineral or hydrocarbon oils, and the "natural" waxes with those of mineral origin.

No attempt has been made to give an exhaustive account of the historical aspect of the various operations and of analytical procedure. If this is desired the reader must consult larger treatises.

The aim, on the contrary, has been to eliminate all matter, the omission of which is compatible with an adequate knowledge of the present-day methods of production, and of analytical control.

The authors have endeavoured to employ as far as possible, a style direct and succinct, and have made use of a special type to indicate at a glance the subject matter of each paragraph. They believe that in this way it has been possible so to minimise the space required as to obtain a volume of convenient size for handling, and adequate in all respects except for academic purposes.

A special feature is the coloured diagrams of the more important analytical determinations. Used in the manner indicated the authors trust that these may prove of value, in particular to the works chemist.

Where possible in the detailed descriptions of individual oils and fats, a series of figures has been given showing (1) the average values (2) the normal variations, and (3) the outside limits recorded for the various analytical data. In the latter case, the authorities responsible for the figures are stated; in the case of the other two series of figures, these have been compiled by a careful comparison of results obtained, in many cases, by the authors, with those published by other chemists.

A companion volume on the practical analytical work referred to in various places in the book is in course of preparation and will shortly be published.

Acknowledgment must in general be made to such works as the now classical treatise on oils and fats by the late Dr Lewkowitsch[1], and to a less extent to other original papers and productions. The figures for the Refractive Indices are mainly those published by Messrs Bolton and Revis[2]. We have to thank Messrs Archbutt and Deeley[3], and their publishers, for permission to print the table of *Viscosities of Glycerine Solutions*. The following firms have kindly placed at our disposal the plates for reproducing many of the illustrations of technical apparatus: Messrs Rose, Downs and Thompson, Messrs Greenwood and Batley, Messrs S. H. Johnson. Our thanks are also due to Mr A. F. Fryer, M.Sc., F.I.C., who kindly consented to finally revise the proofs. For the rest, the authors trust that a practical acquaintance with works processes and of analytical methods extending over a number of years may prove to be of value and service to those engaged in this most interesting and important branch of chemical industry.

PERCIVAL J. FRYER.
FRANK E. WESTON.

RAVENSCAR, TONBRIDGE.
and THE POLYTECHNIC, REGENT ST., W.
January 1917.

[1] *Chemical Technology and Analysis of Oils, Fats and Waxes.* Dr Julius Lewkowitsch. 5th edition. Macmillan. 3 vols.
[2] *Fatty foods.* Bolton and Revis. Churchill.
[3] *Lubrication and Lubricants.* Archbutt and Deeley. C. Griffin and Co.

CONTENTS

SECTION V

SECTION VI

LIST OF PLATES

TABLES OF ANALYTICAL CONSTANTS

available for download from www.cambridge.org/9781107687318

NOTES ON THE USE OF THE TABLES

Each table has a **reference line**, marked off in divisions corresponding to equal fractions of the various values. Opposite to these are placed *sections* hatched in colours to represent the various classes of oils, fats and waxes. The length of these sections indicates the **extreme variations** of each individual oil, etc. The line connecting each section with its corresponding name is placed in the position of the **average value** for each substance.

It is suggested that the tables may be used in the following manner:—In order to identify an unknown oil, fat, or wax, the various values are obtained by analysis, and a rule or straight-edge is placed horizontally at these particular positions on the reference line, when *the sections which appear cut* by the straight-edge represent those oils, etc. which come within possible range of consideration. Further, the coloured line (connecting with the names of the oils, fats, etc.) which is nearest to the straight-edge will indicate the most probable one, since it approaches most closely to the average value for this particular oil, fat, or wax.

Using the tables in succession—commencing preferably with the **iodine value**—and assuming only a single oil, fat, or wax to be present, the correct result can be rapidly obtained, since all other possibilities are automatically eliminated. Thus, if on reference to the table for iodine values the straight-edge intersects say four oils, etc., two of these may be above or below the straight-edge in (for example) the specific gravity table, and of the two remaining possibilities, one may be eliminated on reference to the table of saponification values.

In the case of mixtures of oils, fats and waxes, the problem of identification is a much more complex one, but the tables will still be found helpful in arriving at the correct solution.

In addition to their use for analytical determinations, the tables are also intended to have an educational value to the student. For this purpose, they will repay careful study. When the particular oil, fat or wax is under consideration, its relative position on each of the tables should be noted, and, in this way, a clear conception will readily be obtained of its distinctive features and of its class relationships.

A glance at one of the tables will also serve to show whether that particular test is of service in discriminating a given oil, etc., from any other oil, and thus save the labour of having to refer to the separate descriptions of individual oils.

ERRATA

Table of Specific Gravities (between pp. 68 and 69)—

Cod Liver Oil. Average value should read ·925.

Lard. *For* ·931 *read* ·936 and the hatched bar at 934—·938 (*not* ·931—·932).

Table of Iodine Values (between pp. 80 and 81)—

Neat's Foot Oil, should read 69—72, average value 70.

Palm Kernel. Hatched section omitted should read 10—17.

Table of Bromine Thermal Test (between pp. 82 and 83)—

Neat's Foot Oil should read 12—13, average value 12·7.

SECTION I

INTRODUCTION

§ **1.** THE OILS AND FATS INDUSTRY is essentially a **chemical industry**, i.e., it is based on certain fundamental chemical reactions. In order, therefore, to have an intelligent conception of the nature of the processes involved, and of the classes of materials which are made use of in this industry, it is absolutely essential to possess a knowledge of the **fundamental principles of chemistry** both theoretical and practical. To take only one example out of the hundreds that occur in practice, e.g., the product known technically as "DISTILLED GREASE STEARINE[1]"; in order that one may appreciate the properties of this substance, it is necessary to have a sufficient chemical knowledge to be able to answer, at least, the following questions:

(*a*) What is an acid?
(*b*) What is a base?
(*c*) How are these compounds distinguished?
(*d*) What combinations can each form?
(*e*) What laws govern such combinations?

Many other questions suggest themselves and unless equipped at the outset with the necessary knowledge to answer such queries it will be useless for the student to attempt to grasp the principles underlying works' processes and methods.

§ **2.** Compared however with many other industries, the chemistry of the operations in connection with oils, fats and waxes is remarkable for its simplicity. There are, in fact, three chemical elements only, which are concerned to any extent with the subject. These, in various conditions of combination, comprise all the different species of fatty and waxy bodies. Moreover these three chemical elements are perhaps the commonest and most familiar of all the elements. They are the non-metals :—

Hydrogen	(symbol)	H	(atomic weight)	1
Carbon	,,	C	,,	12
Oxygen	,,	O	,,	16

The atomic weights given are taken to the nearest whole number, and are correct to two places of decimals, rendering chemical calculations very easy.

[1] See § 168.

The **mineral oils and waxes** are, chemically speaking, simpler still, consisting mainly of the two elements CARBON and HYDROGEN in combination.

In the crude state most mineral oils contain small quantities of compounds of **oxygen** and **sulphur**, which are removed during the processes of refining.

§ **3.** The diversity of the compounds present in fatty bodies is attained chemically by the **varying number of carbon atoms** in the molecules of the different compounds, and also by the **different arrangement** of the atoms in the molecules. For example, the substance known as STEARIC ACID, which occurs in animal fats, contains 18 atoms of carbon linked to one another by chemical forces in the form of a chain, and diagrammatically represented thus :—

-C-C-C-C-C-C-C-C-C-C-C-C-C-C-C-C-C-C-[1]

whilst the compound known as Melissic[2] Acid, which occurs in beeswax, contains no less than 30 atoms of carbon linked in a similar manner.

Again, all the acids which occur in oils, fats, etc., contain a certain group of atoms arranged always in the same manner, and known as the "ACID" or "CARBOXYL" group graphically represented as follows:

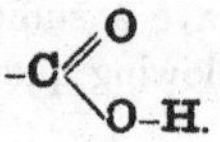

$$-C\begin{matrix} \diagup O \\ \diagdown O\text{-}H. \end{matrix}$$

Further, in all NATURAL OR FIXED OILS AND FATS (i.e., those derived from *animals* or *plants*), without a single exception, the base entering into combination with the numerous fatty-acid radicles, is, strangely enough, the same, viz., the **trihydric basic radicle** of **Glycerine**[3], e.g. $C_3H_5 \equiv$, glycerine being a trihydric alcohol[4], viz., $C_3H_5(OH)_3$.

§ **4.** Oils, fats and waxes consist of the following **natural groups** :

Products of animals and plants	(*a*) Fixed oils, fats and waxes. (*b*) Volatile or Essential oils.
Products of minerals	(*a*) Petroleum[5] oils and waxes. (*b*) Ozokerite[6] or Earth wax. (*c*) Brown coal, bitumen, and peat oils and waxes. (*d*) Shale oils and waxes. (*e*) Coal tar oils.

Bones and rosin on dry distillation give "bone oil" (Dippel's oil) and "rosin oil."

[1] This chain of atoms is not necessarily straight, any more than would be a watch-chain when held in the palm of the hand.

[2] Gk. *melissa* = a bee.

[3] Gk. *glukus* = sweet.

[4] Arabic *al* = the, *Kohl* = stibium = sulphide of antimony.

[5] Lat. *petra* = a rock and *oleum* = oil.

[6] Gk. *ozo* = to smell and *kēros* = wax.

§ **5.** Many plants, as stated above, yield **two kinds of oil.**

(*a*) **Non-volatile**, contained mostly in the *seeds* and *fruits* of plants.

(*b*) **Volatile**, contained mainly in the *leaves, stems* and *flowers* of the plant.

The first group are the so called "**fixed oils.**" They are obtained by means of expression from the seeds, etc., or by extraction with solvents. The second group are known as the "**volatile**" or "**essential oils.**" They are obtained by distillation with steam and are mainly used in perfumery, being all characteristically odorous[1].

§ **6.** Animal and Vegetable oils are in nature produced by a process of **synthesis**, that is, they are built up from simpler substances by means of the activity of the living protoplasm of animal and plant cells. The oil appears as minute spherical globules embedded in the protoplasm of the cell. These globules increase in size and coalesce, frequently constituting the largest portion of the tissue in which they are found.

The **natural function** of oils in plants appears to be to act as a reserve food supply for the young embryo during its development. In animals, the fat forms a warm protective coat around the abdominal viscera and probably also forms a food reserve to some degree.

Beeswax is secreted by the bee as a building material for the honey-comb.

§ **7. Considered chemically**, all fixed oils and fats are compounds of **glycerine** with a **fatty acid**, termed **glycerides**, which form one of the classes of a large number of compounds known as "**esters**" or "**ethereal**[2] **salts.**"

They are recognized by the fact that when they are split up by suitable treatment they yield glycerine and a fatty acid. The **animal and vegetable waxes** are also esters, but the base contained in these is an alcohol *other* than glycerine, and consequently no glycerine is obtainable from them.

§ **8. Mineral oils and waxes** may be divided into the **three groups**:

A. Oils and waxes obtained in a more or less pure condition from the earth, e.g.:

(*a*) Petroleum;
(*b*) Ozokerite.

B. Oils and waxes obtained by extraction of mineral products with suitable solvents, e.g.:

(*a*) Montan Wax from Brown Coal;
(*b*) Bitumen Wax from Bitumen;
(*c*) Peat Wax from Peat.

[1] These form a class of compounds outside the scope of this book and constitute an industry to themselves.

[2] Lat. *æther*=Gk. *aither*=the sky, from Gk. *aitho*=to burn.

C. Oils and waxes obtained only by destructive distillation of mineral products, e.g. :

(*a*) Shale oil and waxes from Shale ;

(*b*) Coal tar oils from Coal tar.

Chemically considered the mineral oils and waxes are mainly mixtures of compounds known as **hydrocarbons**. As previously stated, hydrocarbons are compounds consisting of carbon and hydrogen only, and, as will be shown in a later section, there are many types of hydrocarbons whose differences in properties are mainly due, not to different proportions of carbon and hydrogen, but to the different arrangement of the carbon and hydrogen atoms in the molecule, i.e., to the constitution or structure of the molecule.

SECTION II

CHEMISTRY OF THE OILS, FATS AND WAXES

CHAPTER I

FATTY OILS, FATS AND NATURAL WAXES

§ **9.** Oils and fats are not simple substances. They are substances consisting of more or less definite mixtures of **two or more glycerides**[1], simple or mixed. According to the character of the glycerides they contain—whether solid or liquid, etc.—and the amounts of each, so the properties of the various oils and fats differ one from the other.

§ **10.** Glycerides are **colourless, odourless and tasteless substances**, whilst the oils and fats as obtained in commerce are all more or less distinguished by differences of colour, odour and taste. These properties serve to give in many cases an easily recognisable character to the oils and are produced by small quantities of extraneous substances dissolved in the oil or fat.

§ **11.** A pure glyceride is a compound of **glycerine with a fatty acid**. Consider for a moment the base caustic potash [KOH]; this is capable of combining with an acid to produce a salt and water, e.g. a solution of KOH and a solution of HCl when mixed till the resulting hydrochloric acid solution is neutral, gives a solution of a salt KCl (potassium chloride) and more water H_2O: this reaction is quantitatively represented by the chemical equation

$$\text{Solution of } KOH + \text{solution of } HCl = \text{a solution of } KCl + H_2O.$$

Similarly a solution of KOH can form a solution of a salt when mixed with a solution of a fatty acid, e.g. stearic acid $C_{17}H_{35}.CO.OH$, viz.

$$KOH + C_{17}H_{35}.CO.OH = C_{17}H_{35}.CO.OK + H_2O.$$

Now glycerine is a **trihydric alcohol** and is thus a **triacid base**, so that **one molecule** of glycerine $C_3H_5(OH)_3$ is capable of combining with **three molecules** of a monobasic acid, e.g. a fatty acid. The

[1] One or two fats are known which appear to consist practically of one glyceride only.

reaction between glycerine and stearic acid is represented by the chemical equation[1]

$$\begin{matrix} C_{17}H_{35}.CO.O\boxed{H} \\ C_{17}H_{35}.CO.O\boxed{H} \\ C_{17}H_{35}.CO.O\boxed{H} \end{matrix} + C_3H_5\begin{cases} \boxed{OH} \\ \boxed{OH} \\ \boxed{OH} \end{cases} = \begin{matrix} C_{17}H_{35}.CO.O \\ C_{17}H_{35}.CO.O \\ C_{17}H_{35}.CO.O \end{matrix}\Big\rangle C_3H_5 + 3H.OH.$$

Since glycerine is a triacid base it is theoretically possible for **one molecule** of glycerine to react with either **one, two,** or **three molecules** of a monobasic acid; hence the following types of glycerides are possible with stearic acid, viz.

1. **Mono-glycerides.**

$$\underset{\text{stearic[2] acid}}{\underset{\text{1 molecule of}}{C_{17}H_{35}CO.OH}} + \underset{\text{glycerine}}{\underset{\text{1 molecule of}}{C_3H_5(OH)_3}} = \underset{\text{monostearin}}{\underset{\text{1 molecule of}}{C_{17}H_{35}CO.O.C_3H_5(OH)_2.}} + \underset{\text{water}}{\underset{\text{1 molecule of}}{H_2O}}$$

2. **Di-glycerides.**

$$\underset{\text{stearic acid}}{\underset{\text{2 molecules of}}{2C_{17}H_{35}CO.OH}} + \underset{\text{glycerine}}{\underset{\text{1 molecule of}}{C_3H_5(OH)_3}} = \underset{\text{distearin}}{\underset{\text{1 molecule of}}{(C_{17}H_{35}CO.O)_2 : C_3H_5.OH}} + \underset{\text{water}}{\underset{\text{2 molecules of}}{2H_2O}}$$

3. **Tri-glycerides.**

$$\underset{\text{stearic acid}}{\underset{\text{3 molecules of}}{C_{17}H_{35}CO.OH}} + \underset{\text{of glycerine}}{\underset{\text{1 molecule}}{C_3H_5(OH)_3}} = \underset{\text{tristearin}}{\underset{\text{1 molecule of}}{(C_{17}H_{35}CO.O)_3 : C_3H_5.}} + \underset{\text{water}}{\underset{\text{3 molecules of}}{3H_2O}}$$

The structural or constitutional formula for glycerine is represented by the plane formula

$$\begin{matrix} CH_2.OH\ (1) & \text{or}\ (\alpha) \\ | & \\ CH.OH\ (2) & (\beta) \\ | & \\ CH_2.OH\ (3) & (\gamma) \end{matrix}$$

in which it is seen that the three hydroxyl groups OH are attached to different carbon atoms and are numbered (1), (2) and (3) or (α), (β) and (γ) in order to indicate the position of each group in the molecule. It has been previously stated that the property of a carbon compound depends not only upon its composition by weight of each of the constituent elements but also upon the **structure** or **constitution** of the molecule. On referring to the monoglycerides given above it will be seen that *one* hydroxyl-group of the glycerine molecule has been replaced by *one* acid radicle of the stearic acid molecule and hence at first sight there appear to be *three* possible structures for the monoglyceride, viz.

$$\underset{\text{I}}{C_3H_5\begin{cases} C_{17}H_{35}.CO.O & (\alpha) \\ OH & (\beta) \\ OH & (\gamma) \end{cases}} \qquad \underset{\text{II}}{C_3H_5\begin{cases} OH & (\alpha) \\ C_{17}H_{35}.CO.O & (\beta) \\ OH & (\gamma) \end{cases}}$$

[1] This reaction does not take place completely on mixing the acid and glycerine; the conditions for such reaction are indicated later.

[2] Gk. *stear*=fat.

and

$$C_3H_5 \begin{cases} OH & (\alpha) \\ OH & (\beta) \\ C_{17}H_{35}CO . O & (\gamma) \end{cases}$$

III

Now the positions (α) and (γ) in the glycerine molecule are identical, since on inverting the formula the same structure is obtained, i.e. to say the positions (α) and (γ) are symmetrical about the molecule. Since then, the monoglycerides I and III are identical but are different from II, **two monoglycerides are possible.**

On examining the possible structures of the diglycerides, e.g.

$$C_3H_5 \begin{cases} C_{17}H_{35}CO . O & (\alpha) \\ C_{17}H_{35}CO . O & (\beta) \\ OH & (\gamma) \end{cases} \qquad C_3H_5 \begin{cases} C_{17}H_{35} . CO . O & (\alpha) \\ OH & (\beta) \\ C_{17}H_{35} . CO . O & (\gamma) \end{cases}$$

I II

$$C_3H_5 \begin{cases} OH & (\alpha) \\ C_{17}H_{35} . CO . O & (\beta) \\ C_{17}H_{35} . CO . O & (\gamma) \end{cases}$$

III

it will be seen that structures I and III are identical but different from structure II, hence there are **two possible diglycerides.**

From the above it is also evident that there is only **one possible triglyceride** (in the case of a simple glyceride).

§ **12.** As previously stated the terms "**fatty oils, fats and natural waxes**" are used to designate those oils, fats and waxes which are the products of animals and plants as distinguished from those oils and waxes which result from the treatment of certain mineral products.

All fatty oils and solid fats yield glycerine on suitable treatment. This capacity of yielding glycerine is the ultimate test for distinguishing between fatty oils or fats and those substances which in many respects are similar to them.

§ **13. The natural waxes** (i.e. those derived from animal and vegetable sources) resemble fats very closely in most of their properties. They, however, contain no glycerides and thus can yield no glycerine on being subjected to the same treatment that causes fats to yield glycerine.

The popular names for some of these substances are misnomers, e.g. **Japan wax** and **myrtle wax** are both **true fats**, yielding glycerine when split up, whilst **sperm oil** is not a fatty oil but a liquid wax, since no glycerine is obtainable from it.

The natural waxes, however, agree with the fatty oils, in this respect, that they are compounds of fatty acids with alcohols, i.e. they are esters; the alcohols obtained from waxes being monohydric. whilst glycerine is trihydric.

§ **14. Glycerides**, on treatment with water in the form of high pressure steam, are split up or decomposed into two parts, which, combining with the elements of water, produce a **fatty acid** and **glycerine.** Taking, for example, the triglyceride stearin, the reaction is represented by the following chemical equation :

$$\left.\begin{array}{l}C_{17}H_{35}.CO.O\\C_{17}H_{35}.CO.O\\C_{17}H_{35}.CO.O\end{array}\right\rangle C_3H_5 + \begin{array}{l}H.OH\\H.OH\\H.OH\end{array} = \begin{array}{l}C_{17}H_{35}CO.OH\\C_{17}H_{35}CO.OH\\C_{17}H_{35}CO.OH\end{array} + C_3H_5\left\langle\begin{array}{l}OH\\OH\\OH\end{array}\right.$$

1 molecule of tristearin | 3 molecules of water | 3 molecules of stearic acid | 1 molecule of glycerine

or written in contracted form,

$$(R.CO.O)_3 : C_3H_5 + 3H_2O = 3R.CO.OH + C_3H_5(OH)_3$$

where R represents a monovalent hydrocarbon radicle, known as an **alkyl group.**

It will thus be observed that it is incorrect to speak of a fat *containing* glycerine since this body is only produced on the fat being *split up* and the glyceryl radicle $C_3H_5\equiv$ combining with three $-OH$ radicles of water.

From the foregoing equation it will be seen that one molecule of a triglyceride on being split up combines with the elements of water producing **one molecule** of **glycerine** and **three molecules** of a **fatty acid.**

In the case of **waxes** which are decomposed on treatment with a strong base, such as caustic soda or caustic potash, only **one molecule** of a **fatty acid** or one molecule of a salt of the fatty acid is obtained for one molecule of the ester split up. Thus Spermaceti[1] consists largely of the ester composed of Palmitic acid and Cetyl alcohol $C_{15}H_{31}CO.O.C_{16}H_{33}$, and is decomposed by caustic potash as represented by the following equation

$$C_{15}H_{31}.CO.O.C_{16}H_{33} + K.OH = C_{15}H_{31}.CO.OK + C_{16}H_{33}OH$$

Cetyl palmitate | Potassium hydrate | Potassium palmitate | Cetyl alcohol

i.e. **one molecule** of the ester yields **one molecule** of a salt of a fatty acid.

§ **15. Carnaüba wax** contains an ester, amongst others, composed of **two molecules** of a fatty acid combined with one molecule of a dihydric alcohol, which it yields on treatment with caustic potash ; such alcohols are termed glycols, the one obtained from Carnaüba wax having the formula $C_{25}H_{50}(OH)_2$.

[1] Gk. *sperma*=seed, *ketos*=a whale.

§ **16.** It has already been shewn that there are three classes of glycerides, viz.

A. Monoglycerides of general formula $R.CO.O.C_3H_5(OH)_2$,
B. Diglycerides „ „ $(R.CO.O)_2C_3H_5.OH$,
C. Triglycerides „ „ $(R.CO.O)_3C_3H_5$,

where R represents an alkyl group (see § 14).

Apparently **only the triglycerides occur in nature.**

Reimer and Will found in an old sample of rape-oil the **diglyceride** of Erucic[1] acid, viz. $(C_{21}H_{41}.CO.O)_2C_3H_5.OH$. This however was probably formed by the loss of a molecule of Erucic acid from the triglyceride, on rancidity taking place.

§ **17.** A. **Monoglycerides.** These were first obtained synthetically by Berthelot who heated fatty acids with excess of glycerine in a sealed tube; both mono- and diglycerides were obtained.

The modern method is to mix equivalent proportions of mono-chlor-hydrin and finely powdered sodium salts of the fatty acids and heat. Sodium chloride separates and the glyceride is extracted with ether and filtered through charcoal. The reaction that takes place is represented thus:

$$\begin{array}{llll} CH_2.OH & & CH_2.OH & \\ | & & | & \\ CH.OH + & R.CO.O.Na = & CH.OH & + NaCl \\ | & & | & \\ CH_2.Cl & & CH_2.R.CO.O & \\ \text{mono-chlor-hydrin} & \text{sodium salt of fatty acid} & \text{mono-glyceride} & \text{sodium chloride} \end{array}$$

As before stated (see § 11) there are **two possible** monoglycerides of a fatty acid. The following are a few of the monoglycerides:

(i) α Monoformin[2] $C_3H_5\begin{cases} H.CO.O\ \alpha \\ (OH)_2 \end{cases}$ B.P. in vacuo 165° C.

(ii) α Monoacetin[3] $C_3H_5\begin{cases} CH_3.CO.O\ \alpha \\ (OH)_2 \end{cases}$ a thick liquid, soluble in water: B.P. at 2 mm. 130° C.; S.G. at 15° C. 1·2212.

(iii) α Monobutyrin[4] $C_3H_5\begin{cases} C_3H_7.CO.O\ \alpha \\ (OH)_2 \end{cases}$ an oily liquid; B.P. 270° C. S.G. at 17° C. 1·008; 8 volumes of ester mix with three volumes of water.

(iv) α Monolaurin[5] $C_3H_5\begin{cases} C_{11}H_{23}.CO.O\ \alpha \\ (OH)_2 \end{cases}$ M.P. 59° C.

(v) α Monomyristin[6] $C_3H_5\begin{cases} C_{13}H_{27}CO.O\ \alpha \\ (OH)_2 \end{cases}$ M.P. 68° C.

[1] Gk. *ereugomai* = to vomit. Erucic is derived from the Eruca a genus of Cruciferae whose seeds applied to the skin produce blisters.
[2] Lat. *forma* = an ant.
[3] Lat. *acetum* = vinegar.
[4] Lat. *butyrum* = butter.
[5] Lat. *laurus* = laurel.
[6] Lat. *myristica* = a nutmeg.

(vi) α Monopalmitin $C_3H_5\begin{cases} C_{15}H_{31}.CO.O\ \alpha \\ (OH)_2 \end{cases}$ M.P. 72° C. soluble to extent of 5 % in absolute alcohol.

(vii) α Monostearin $C_3H_5\begin{cases} C_{17}H_{35}CO.O\ \alpha \\ (OH)_2 \end{cases}$ M.P. 73° C.; crystallizes in microscopic needles.

(viii) α Monocerotin[1] $C_3H_5\begin{cases} C_{25}H_{51}.CO.O\ \alpha \\ (OH)_2 \end{cases}$ M.P. 78·8° C.

(ix) α Monomelissin $C_3H_5\begin{cases} C_{29}H_{59}CO.O\ \alpha \\ (OH)_2 \end{cases}$ M.P. 92° C.

(x) α Monoolein[2] $C_3H_5\begin{cases} C_{17}H_{33}CO.O\ \alpha \\ (OH)_2 \end{cases}$ a yellowish liquid: S.G. at 21° C. 0·947; solidifies at 0° C.

Gruss[3] has recently prepared β monoglycerides by using dichlorhydrin and removing chloro-derivatives by the action of silver nitrite.

(xi) β Monolaurin $C_3H_5\begin{cases} OH & \alpha \\ C_{11}H_{23}.CO.O & \beta \\ OH & \gamma \end{cases}$ small white crystals. M.P. 61° C.

(xii) β Monopalmitin $C_3H_5\begin{cases} OH & \alpha \\ C_{15}H_{31}.CO.O & \beta \\ OH & \gamma \end{cases}$ white leaflets. M.P. 74° C.

B. **Diglycerides.** These are prepared by similar methods to the foregoing, viz. the heating of a mixture of one molecule of dichlorhydrin and two molecules of the sodium salt of a fatty acid. There are two possible structural forms of a diglyceride (see § 11), viz.:

$$C_3H_5\begin{cases} R.CO.O\ \alpha \\ OH \\ R.CO.O\ \alpha \end{cases} \quad + \quad C_3H_5\begin{cases} R.CO.O\ \alpha \\ R.CO.O\ \beta \\ OH \end{cases}$$

termed αα or symmetrical diglycerides and αβ or asymmetric diglycerides.

The following are some of those which have been prepared:

(i) Diformin $C_3H_5\begin{cases} (H.CO.O)_2 \\ OH \end{cases}$ prepared commercially by heating 10 parts of pure formic acid H.CO.OH with 4 parts of 95 % glycerine to 140° C. Dilute formic acid distils over and diformin is left.

S.G. at 15° C. 1·304. B.P. at 20–30 mm. 163°–166° C.

(ii) Diacetin $C_3H_5\begin{cases} (CH_3CO.O)_2 \\ OH \end{cases}$ formed together with monoacetin on heating anhydrous glycerine with acetic acid. It is a commercial product and used to adulterate essential oils; S.G. at 15° 1·178; B.P. at 40 mm. 175° C.

(iii) αα Dibutyrin $C_3H_5\begin{cases} C_3H_7.CO.O\ \alpha \\ OH \\ C_3H_7.CO.O\ \alpha \end{cases}$ S.G. at 17° 1·083 B.P. at 19 mm. 173–176° C.

[1] Lat. *cerotum* = wax. [2] Lat. *oleum* = oil.
[3] *Ber.* 1910. 43. 1288, 1291.

(iv) *αβ* Dibutyrin $C_3H_5\begin{cases} C_3H_7\,CO\,.\,O\ \alpha \\ C_3H_7\,.\,CO\,.\,O\ \beta \\ OH \end{cases}$ B.P. at 19 mm. 166°–168° C.

(v) *αα* Dilaurin $C_3H_5\begin{cases} C_{11}H_{23}\,.\,CO\,.\,O\ \alpha \\ OH \\ C_{11}H_{23}\,.\,CO\,.\,O\ \alpha \end{cases}$ M.P. 57° C.

(vi) *αα* Dimyristin $C_3H_5\begin{cases} C_{13}H_{27}\,.\,CO\,.\,O\ \alpha \\ OH \\ C_{13}H_{27}\,.\,CO\,.\,O\ \alpha \end{cases}$ M.P. 61° C.

(vii) *αα* Dipalmitin $C_3H_5\begin{cases} C_{15}H_{31}\,.\,CO\,.\,O\ \alpha \\ OH \\ C_{15}H_{31}\,.\,CO\,.\,O\ \alpha \end{cases}$ M.P. 70° C.

(viii) *αβ* Dipalmitin $C_3H_5\begin{cases} C_{15}H_{31}\,.\,CO\,.\,O\ \alpha \\ C_{15}H_{31}\,.\,CO\,.\,O\ \beta \\ OH \end{cases}$ M.P. 67° C.

(ix) *αα* Distearin $C_3H_5\begin{cases} C_{17}H_{35}\,.\,CO\,.\,O\ \alpha \\ OH\ \beta \\ C_{17}H_{35}\,.\,CO\,.\,O\ \alpha \end{cases}$ M.P. 72·5° C.

(x) *αβ* Distearin $C_3H_5\begin{cases} C_{17}H_{35}\,.\,CO\,.\,O\ \alpha \\ C_{17}H_{35}\,.\,CO\,.\,O\ \beta \\ OH\ \alpha \end{cases}$ M.P. 74·5° C.

(xi) *αα* Diarachin[1] $C_3H_5\begin{cases} C_{19}H_{39}\,.\,CO\,.\,O\ \alpha \\ OH \\ C_{19}H_{39}\,.\,CO\,.\,O\ \alpha \end{cases}$ M.P. 75° C.

(xii) Dicerotin $C_3H_5\begin{cases} (C_{25}H_{51}\,.\,CO\,.\,O)_2 \\ OH \end{cases}$ M.P. 79·5° C.

(xiii) Dimelissin $C_3H_5\begin{cases} (C_{29}H_{59}\,.\,CO\,.\,O)_2 \\ OH \end{cases}$ M.P. 90° C.

(xiv) *αα* Diolein $C_3H_5\begin{cases} C_{17}H_{33}\,.\,CO\,.\,O\ \alpha \\ OH \\ C_{17}H_{33}\,.\,CO\,.\,O\ \alpha \end{cases}$ Oily liquid solidifies at 0° C. to a white mass.

(xv) Dierucin $C_3H_5\begin{cases} (C_{21}H_{41}\,.\,CO\,.\,O)_2 \\ OH \end{cases}$ M.P. 47° C. as stated § 16 occurs in old rape oil.

C. **Triglycerides.** These form 98 to 99 per cent. of almost all neutral fats and fatty oils. There are two classes of triglycerides, viz. :

(i) **Simple.** Those in which the three acid radicles of the ester are the same, thus yielding only one fatty acid on splitting, e.g.

Tristearin $C_3H_5\begin{cases} C_{17}H_{35}\,.\,CO\,.\,O \\ C_{17}H_{35}\,.\,CO\,.\,O \\ C_{17}H_{35}\,.\,CO\,.\,O \end{cases}$ or

generally $C_3H_5\ (R\,.\,CO\,.\,O)_3$.

[1] Gk. *arachos*, the name of a leguminous plant. Applied to the oil obtained from the earth-nut, *Arachis hypogæa*.

(ii) **Mixed.** Those in which the acid radicles of the ester are different, thus yielding two or three different fatty acids on splitting, e.g.

$$C_3H_5\begin{cases}(R.CO.O)_2\\R'.CO.O\end{cases} \qquad C_3H_5\begin{cases}R.CO.O\\R'.CO.O\\R''.CO.O,\end{cases}$$

where R, R' and R'' are different alkyl groups.

Up to the last few years, fats were considered to be **mixtures of simple glycerides.** Now however many mixed triglycerides have been isolated from fats and oils and their occurrence is probably very widespread.

The isolation of these mixed triglycerides has been accomplished by the fractional crystallization of the glycerides from oils and fats. The isolation of the pure triglyceride is an exceedingly laborious task, e.g. Duffy, an early experimenter, crystallized 2000 grams of mutton tallow from ether *thirty-two successive times*, obtaining finally only 8 grams of a substance which even then was not pure.

A curious property of triglycerides is the so-called "**double-melting point**" which they exhibit. Thus the glyceride tristearin shortly after fusion and resolidification melts at 55° C., but some time after resolidification the melting point is 71° C., which it retains until again being remelted and solidified.

This phenomenon is generally attributed to the existence of two modifications of the substance, one termed the **unstable** or **labile** form which passes into the higher or "**stabile**" modification. Much uncertainty, however, still exists on this question[1].

The following list (p. 13) includes the simple triglycerides which have been mostly prepared by the interaction of glycerine and the fatty acid.

It will be noticed that with the increase of molecular weight, the B.P. and M.P. rise, whilst the S.G. decreases, and the glycerides become less and less soluble in solvents, such as water, alcohol and ether, in the order named. The first fourteen of the glycerides mentioned in the table form a series of compounds known as **saturated** fats, which will be explained later when the fatty acids are considered.

Olein, erucin, ricinolein, linolein, linolenein and clupanodonin are **unsaturated** fats.

Ricinolein, the characteristic constituent of castor oil, behaves differently from all the other glycerides in being soluble in alcohol. This is due to the presence of an hydroxyl group (OH) attached to one of the carbon atoms in the chain.

§ **18. Mixed Triglycerides** in which there are two acid groups present are possible in **two modifications,** e.g.:

$$C_3H_5\begin{cases}R.CO.O\\R^1.CO.O\\R.CO.O\end{cases} \text{ and } C_3H_5\begin{cases}R.CO.O\\R.CO.O\\R^1.CO.O\end{cases}$$

[1] See le Chatelier, *Compt. Rend.* 1913, **156**, 589. Grün, *Berichte*, 1912, **45**, 3691.

Triglyceride	Formula	Molecular weight	No. of C atoms in fatty acid	Specific gravity	B. P.	M. P.	Solubility	Natural occurrence
Triformin	$C_3H_5(H.CO.O)_3$	176	C_1	1·4412/18°	266°	—	—	Not in nature
Triacetin	$C_3H_5(CH_3.CO.O)_3$	218	C_2	1·1603/15°	258°	—	Water 7 % miscible alcohol	Seeds of Euonymus Europea?
Tributyrin	$C_3H_5(C_3H_7.CO.O)_3$	302	C_4	1·0324/$\frac{20^{\circ}}{4}$	287°	—	Water nearly insoluble	Butter, mixed glyceride
Trivalerin[1]	$C_3H_5(C_4H_9.CO.O)_3$	344	C_5	—	—	—	Sol. alcohol	Porpoise and dolphin oils?
Tricaproin[2]	$C_3H_5(C_5H_{11}.CO.O)_3$	386	C_6	·9817/$\frac{20^{\circ}}{4}$	—	− 25°	,, ,,	Butter, cocoa-nut and palm-nut oil as mixed glyceride
Tricaprylin[2]	$C_3H_5(C_7H_{15}.CO.O)_3$	470	C_8	·9540/$\frac{20^{\circ}}{4}$	—	+ 8°	,, ,,	,, ,,
Tricaprin[2]	$C_3H_5(C_9H_{19}.CO.O)_3$	554	C_{10}	·9205/$\frac{40^{\circ}}{4}$	—	31°	Sparingly alcohol	,, ,,
Trilaurin	$C_3H_5(C_{11}H_{23}.CO.O)_3$	638	C_{12}	·8944/$\frac{60^{\circ}}{4}$	—	45°	,, ,,	Laurel oil
Trimyristin	$C_3H_5(C_{13}H_{27}.CO.O)_3$	722	C_{14}	·8848/$\frac{60^{\circ}}{4}$	—	56°	,, ,,	Nutmeg butter
Tripalmitin	$C_3H_5(C_{15}H_{31}.CO.O)_3$	806	C_{16}	·8657/$\frac{80^{\circ}}{4}$	—	63°	Very sparingly alc.	Japan wax
Tristearin	$C_3H_5(C_{17}H_{35}.CO.O)_3$	890	C_{18}	·8621/$\frac{80^{\circ}}{4}$	—	71·6°	,, ,, ,,	Most solid fats
Triarachin	$C_3H_5(C_{19}H_{39}.CO.O)_3$	974	C_{20}	—	—	—	Slightly sol. ether	Arachis oil
Tricerotin	$C_3H_5(C_{25}H_{51}.CO.O)_3$	1226	C_{26}	—	—	77°	—	Dandelion oil
Trimelissin	$C_3H_5(C_{29}H_{59}.CO.O)_3$	1394	C_{30}	—	—	89°	—	Not in nature
Triolein	$C_3H_5(C_{17}H_{33}.CO.O)_3$	884	C_{18}	·900/15°	—	S.P. − 4°	Slightly sol. alcohol	Non-drying oils?
Trierucin	$C_3H_5(C_{21}H_{41}.CO.O)_3$	1052	C_{22}	—	—	31°	Nearly insol. alc.	Rape oil class
Triricinolein[3]	$C_3H_5(C_{17}H_{32}.OH.CO.O)_3$	932	C_{18}	·959/15°	—	—	Sol. 96 % alcohol	Castor oil class
Trilinolein[4]	$C_3H_5(C_{17}H_{31}.CO.O)_3$	878	C_{18}	—	—	—	—	Vegetable drying oils
Trilinolenein	$C_3H_5(C_{17}H_{29}.CO.O)_3$	872	C_{18}	—	—	—	—	Linseed oil
Triclupanodonin[5]	$C_3H_5(C_{17}H_{27}.CO.O)_3$	866	C_{18}	—	—	—	—	Fish oils

[1] Lat. *valeo* = to be strong. Valerian plant so called because of its powerful medicinal properties.
[2] Lat. *capra* = a she goat.
[3] Lat. *ricinus* = a tick. *Ricinus communis*, the point of which was supposed to resemble a tick.
[4] Lat. *linum* = flax.
[5] Lat. *clupea* = a fish.

whilst there are **three possible forms** where all the acid groups are different, e.g.:

$$C_3H_5\begin{cases}R^1.CO.O\\R^2.CO.O\\R^3.CO.O\end{cases}\quad C_3H_5\begin{cases}R^1.CO.O\\R^3.CO.O\\R^2.CO.O\end{cases}\text{ and }C_3H_5\begin{cases}R^2.CO.O\\R^1.CO.O\\R^3.CO.O\end{cases}$$

the three positions of the acid groups being designated as before α, β, and γ.

The following mixed triglycerides are amongst those which have been isolated:

(i) α Palmito-$\beta\gamma$ distearin $C_3H_5\begin{cases}C_{15}H_{31}.CO.O\ \alpha\\(C_{17}H_{35}.CO.O)_2\ \beta\gamma\end{cases}$ isolated from lard by Bömer[1], M.P. 68°.

(ii) a Stearo-dipalmitin $C_3H_5\begin{cases}C_{17}H_{35}.CO.O\\(C_{15}H_{31}.CO.O)_2\end{cases}$ also obtained by Hansen, M.P. 55° C.; it was probably impure.

(iii) a Myristo-palmito-olein $C_3H_5\begin{cases}C_{13}H_{27}.CO.O\\C_{15}H_{33}.CO.O\\C_{17}H_{33}.CO.O\end{cases}$ stated by Klimont to occur in cocoa-butter. M.P. 25° C.

(iv) an Oleo-dipalmitin $C_3H_5\begin{cases}C_{17}H_{33}.CO.O\\(C_{15}H_{31}.CO.O)_2\end{cases}$ has been obtained from tallow and said to melt at 48° C. (Hansen).

(v) Oleo-distearin $C_3H_5\begin{cases}C_{17}H_{33}.CO.O\\(C_{17}H_{35}.CO.O)_2.\end{cases}$ M.P. 44° C.

This was the first known mixed glyceride and was obtained by Heise from mkanyi fat and from cacao-butter by precipitating ethereal solutions of these with alcohol.

An important body, which occurs in maize oil, viz. **Lecithin**, is actually a triglyceride, containing several fatty acid radicles and one radicle of a substituted phosphoric acid; its general formula can be represented thus:

$$C_3H_5\begin{cases}R^1.CO.O\\R^2.CO.O\\OX\end{cases}$$

where R^1 and R^2 represent alkyls as before and X represents the complex group:—

$$-P\begin{cases}OH\\=O\\O.CH_2.CH_2\end{cases}\!\!\!\!\begin{matrix}\\ \\ \searrow N(CH_3)_3.\\ HO\nearrow\end{matrix}$$

The fatty acids obtained on splitting Lecithin from egg yolk are palmitic, stearic and linolic so that it seems that Lecithin is a very complex mixed triglyceride. It is optically active and belongs to the large group of recently discovered "Phosphatides."

§ 19. **The glycerides most frequently found** in oils and fats are tripalmitin, tristearin and triolein, although **mixed glycerides** with these acids as radicles undoubtedly occur in many cases.

[1] Bömer suggests that the palmito-distearin he obtained from mutton tallow is the β compound

$$C_3H_5\begin{cases}C_{17}H_{35}.CO.O\\C_{15}H_{31}.CO.O\ \beta\\C_{17}H_{35}.CO.O\end{cases}$$

In addition to these glycerides, all oils and fats contain small quantities of other substances, due to the methods of extraction from the seeds, etc. Some of these can be removed by steaming and washing or filtration. There still remain however traces of bodies which, in many cases, give a distinctive colour, smell or flavour to the oil. Besides these bodies, which may be regarded as impurities, there also occur substances which must be regarded as natural constituents of oils and fats. The most important of these are two complex alcohols known as **phytosterol**[1] and **cholesterol**[2]; cholesterol has the composition $C_{27}H_{45}.OH$, being a monohydric alcohol.

§ **20.** All oils contain more or less **solid glycerides**, which, on slow cooling, separate out, giving the so-called "*stearines*" of commerce. The oil is then said to be "demargarinated"[3] (edible oils), or "racked" (cod-liver oil).

§ **21.** The **solubility** of oils in various solvents is an important question in commerce. They may be considered as completely insoluble in pure water, though traces are dissolved on shaking. In this operation, the oil is split up into minute globules giving an apparently homogeneous liquid of a creamy consistence, termed an **emulsion.**

This emulsion gradually separates on standing, the particles of oil rising to the surface and coalescing, giving a clear liquid once more. Certain substances greatly assist the formation of emulsions and tend to give a permanence to them.

Oils and fats are sparingly soluble in **cold alcohol** with one notable exception—**castor oil**[4].

They dissolve readily in ETHER, CHLOROFORM, CARBON TETRACHLORIDE, CARBON DISULPHIDE, BENZENE, PARAFFIN OILS, and PETROLEUM ETHER. The exception to this rule is again CASTOR OIL which is only sparingly soluble in the last two solvents. Oils and fats dissolve small quantities of SULPHUR and PHOSPHORUS at ordinary temperature, a fact which has some technical importance.

§ **22.** On **heating** above 250° some oils, notably the drying oils, thicken, a change probably due to "Polymerisation[5]," whilst tung oil is converted into a solid mass. On continuing the heating all oils begin to decompose, giving off volatile products, of which the most important is the substance known as "acrolein" $C_2H_3.CHO$, a derivative of glycerine, recognized by its penetrating acrid odour. Hydrocarbons are also produced in quantities dependent on the degree of heat, and especially when distilling under pressure.

[1] Gr. *phuton* = a plant and *stear* = fat.
[2] Gr. *cholē* = the bile and *stear* = fat. [3] Lat. *margarita* = a pearl.
[4] Lat. *castor* = a beaver. Castoreum is the name for a mucilaginous substance found in the two inguinal sacs of the beaver. Castor oil so called because of its supposed resemblance to castoreum. [5] See § 29.

§ **23.** The effect of the **atmosphere** on oils and fats is a complex one, and is best treated by considering the action of the three agents **light**, **air** and **moisture**.

The action of **light** alone on dry oil protected from the air is chiefly noticed in its **bleaching properties**. This fact is made use of in the case of certain oils for special purposes, e.g. artists' oils. The action of light in bleaching beeswax is also of commercial importance. Light alone is unable to produce rancidity in oils and fats.

The action of **air** apart from light and moisture on dry oils is practically nil. In the presence of even **traces of moisture** the effect varies greatly according to the chemical composition of the oil. Those oils known as **drying oils** absorb oxygen from the air and are converted finally into solid substances. Other oils absorb oxygen to a greater or less extent. The subject is fully discussed in a later chapter.

The effect of **moisture** on an oil or fat is generally speaking to produce a greater or less degree of **rancidity**. This change is accompanied by the formation of free fatty acids due to the decomposition of the oil (see § 3). The natural sweetness of the oil is lost and it acquires a disagreeable odour and flavour—a result not entirely to be attributed to the presence of free fatty acids.

The decomposition of oils and fats with formation of free fatty acids when exposed to the air, which always is more or less moist, has now been proved to be greatly accelerated by the presence of soluble ferments, or **enzymes**. These are probably contained in all seeds and small amounts pass into the oil on expression or extraction. The action of enzymes in splitting oils has attained commercial importance. The ferment from castor seed is employed for this purpose.

Rancidity however seems to be due, not entirely or even chiefly to the presence of the free acids formed as described above. It would appear to be also the result of the further action of air and moisture on the fatty acids thus formed and probably also on the glycerine set free. Amongst these secondary products are the lower volatile fatty acids and certain aldehydes, especially œnanthaldehyde $C_6H_{13}.CHO$.

§ **24.** **Hydrogen** gas has no action on oils and fats. In the presence, however, of certain finely divided metals the oil is hardened to a solid fat. This reaction, which is now the basis of an important commercial process, is discussed in all its bearings in a subsequent chapter. The metals chiefly used are nickel and palladium. Their action in helping to produce the change is known as **catalytic**, a phenomenon which is familiar in many other branches of chemical industry.

§ **25.** **Other reagents** act in different ways on oils and fats, viz.:

(i) **Chlorine and Bromine** are absorbed with evolution of hydrochloric and hydrobromic acids respectively except in the case of unsaturated esters. Their action is further discussed later on (see Iodine Value).

(ii) **Iodine** is slowly dissolved by oils, the amount depending on the nature of the oil.

(iii) **Sulphur Chloride** readily attacks oils and fats forming so-called "**Vulcanized fats.**"

(iv) **Nitric Acid** in concentrated form acts energetically on oils and fats with the evolution of oxides of nitrogen in the form of red fumes.

(v) **Nitrous Acid** affects the constituent **triolein** in oils and fats, converting this into a solid body known as **trielaïdin.** Thus non-drying oils, containing most triolein are solidified on treating with nitrous acid.

(vi) **Concentrated Sulphuric Acid** produces darkening of the oil or fat with a rise of temperature and evolution of sulphur dioxide SO_2. If kept cold and mixed gradually the oil is converted into glycerides of a complex nature. At temperatures above 100° C. the oil is partially split up into glycerine and sulpho-compounds of the fatty acids.

(vii) **Alkalies** split up the glycerides in the oil, combining with the fatty acid radicle to form a salt known as a **Soap** and liberate the glycerine. This reaction forms the basis of the soap industry (see § 13).

CHAPTER II

MINERAL OILS AND WAXES

§ **26.** Petroleum and crude paraffin, i.e., the crude oil obtained by the distillation of carbonaceous material, e.g., shale, are complex mixtures of **hydrocarbons**; crude paraffin also contains carbon compounds of *nitrogen* and *oxygen*, whilst some petroleums contain sulphur compounds.

§ **27.** Though, as previously stated, hydrocarbons are compounds containing only **carbon** and **hydrogen** (§§ 2, 8), great diversity in properties is exhibited amongst the many members that form this class of compounds. It has been found possible, however, to classify the immense number of hydrocarbons into well-defined **groups** or **series**, the members of each series, though differing amongst themselves, possessing some property or properties characteristic of that series which distinguish them from the members of another series. The members of a series of hydrocarbons are characterized by:

(1) possessing the same **structure**;

(2) possessing similar properties; and

(3) each member differing in composition from the next more complex member by **one atom of carbon** and **two atoms of hydrogen**.

§ **28.** Members of the following series occur in petroleum or paraffin or in both.

I. **Paraffins or methane hydrocarbons.**

The members of this series contain carbon atoms which are singly linked to each other, i.e., each member is a **saturated** carbon compound (§ 39).

The first few members of this series are:

CH_4 or $C_1H_{(2\times1)+2}$ Methane
C_2H_6 or $C_2H_{(2\times2)+2}$ Ethane
C_3H_8 or $C_3H_{(2\times3)+2}$ Propane
C_4H_{10} or $C_4H_{(2\times4)+2}$ Butane
C_5H_{12} or $C_5H_{(2\times5)+2}$ Pentane

Paraffins

Name	Formula	M.P.	B.P.		S.G.	Occurrence
Methane	CH_4	– 186°	– 164°		·415 @ B.P.	Natural gas; coal gas; fire damp; marsh gas
Ethane	C_2H_6	– 172°	– 84°, 4° C. (46 atmos.)		·446 @ 0°	Natural gas and petroleum
Propane	C_3H_8	—	– 37° C.		·536 @ 0°	Petroleum
Butanes						
Normal	C_4H_{10}	(below 0°)	1° C.		·600 @ 0°	Petroleum
Iso-	$CH:(CH_3)_3$	—	– 17° C.		—	Petroleum
Pentanes						
Normal	C_5H_{12}	—	37°—39°		·626 @ 17° C.	Petroleum, petroleum ether
Iso-	$C_2H_5.CH:(CH_3)_2$	—	30°		·638 @ 14° C.	Petroleum, petroleum ether
Tertiary	$(CH_3)_2:C:(CH_3)_2$	– 20°	9·5°		—	—
Hexanes						
Normal	C_6H_{14}	—	71°		·663 @ 17° C.	Petroleum, petroleum ether
β Methylpropane	$C_3H_5.CH:(CH_3)_2$	—	62°		—	—
ββ Dimethylbutane	$(CH_3)_2.CH.CH:(CH_3)_2$	—	58°		—	Baku petroleum ether
β Ethyl-butane	$CH_3.CH:(C_2H_5)_2$	—	?		—	—
β Dimethylbutane	$(CH_3)_3:C.C_2H_5$	—	43°—48°		—	—
Heptane n.	C_7H_{16}	—	99°		·6967 @ 19°	Petroleum, tar oil from cannel coal
Octane n.	C_8H_{18}	—	125°		·718 @ 0°	and ligroine
Nonane n.	C_9H_{20}	– 51°	149·5°	@ 760 mm.	·733 @ 0°	Petroleum, tar oils from distillation of turf, lignite, bituminous coal and shale
Decane n.	$C_{10}H_{22}$	– 32°	173°		·7456 @ 0° C.	,, ,,
Undecane n.	$C_{11}H_{24}$	– 26·5°	194·5°		·7745 @ M.P.	,, ,,
Dodecane n.	$C_{12}H_{26}$	– 12°	214°		·773 @ M.P.	,, ,,
Tridecane n.	$C_{13}H_{28}$	– 6·2°	234°		·775 @ M.P.	,, ,,
Tetradecane n.	$C_{14}H_{30}$	4·5°	252·5°		·775 @ M.P.	,, ,,
Pentadecane n.	$C_{15}H_{32}$	10°	270·5°		·775 @ M.P.	,, ,,
Hexadecane	$C_{16}H_{34}$	18°	287·5°		·775 @ M.P.	,, ,,
Heptadecane	$C_{17}H_{36}$	22·5°	303°		·776 @ M.P.	,, ,, and scaly paraffin
Octodecane	$C_{18}H_{38}$	28°	317°		·776 @ M.P.	,, ,,
Nondecane	$C_{19}H_{40}$	32°	330°		·777 @ M.P.	,, ,, and vaselines
Eicosane	$C_{20}H_{42}$	36·7°	205°		·778 @ M.P.	,, ,, and vaselines
Heneicosane	$C_{21}H_{44}$	40·4°	215°		·778 @ M.P.	,, ,, and solid paraffin wax,
Docosane	$C_{22}H_{46}$	44·4°	244·5°		·778 @ M.P.	ozokerite
Tricosane n.	$C_{23}H_{48}$	47·7°	234°	@ 15 mm.	·778 @ M.P.	—
Tetracosane n.	$C_{24}H_{50}$	51·1°	243°		·778 @ M.P.	Paraffin wax
Heptacosane n.	$C_{27}H_{56}$	59·5°	270°		·779 @ M.P.	—
Hentriacontane n.	$C_{31}H_{64}$	68·1°	302°		·780 @ M.P.	—
Dotriacontane	$C_{32}H_{66}$	70·0°	310°		·781 @ M.P.	—
Pentatriacontane	$C_{35}H_{72}$	74·7°	331°		·782 @ M.P.	—
Hexacontane	$C_{60}H_{122}$	102°	—		—	—

or writing a general formula which represents any member of the series

$$C_nH_{2n+2},$$

where *n* represents any whole number.

The constitution or structure of these bodies is similar. Taking C_3H_8 as an example, it is

```
   H  H  H
   |  |  |
H—C—C—C—H        or in a contracted form CH3 . CH2 . CH3,¹
   |  |  |
   H  H  H
```

again, C_5H_{12} has the structure

```
   H  H  H  H  H
   |  |  |  |  |
H—C—C—C—C—C—H        or  CH3 . CH2 . CH2 . CH2 . CH3.
   |  |  |  |  |
   H  H  H  H  H
```

It will be noted that no carbon atom is united to another carbon atom by more than *one* force or valency and that each carbon atom is united to *two* hydrogen atoms, except the two end ones which are joined to three; each C atom is therefore fully exercising its **tetra-valent** function by being in combination with four distinct atoms. All paraffins possessing the above structure are termed **normal** paraffins.

Now, if the structure of butane be examined, it will be found possible to arrange the carbon atoms in *two* distinct ways, but with this reservation that no carbon atom is to be united to another carbon atom by more than one force, viz.:

```
                                           H     H
   H  H  H  H                  H            \   /
   |  |  |  |                  |             >C<
H—C—C—C—C—H       and    H—C—C—C           \
   |  |  |  |                  |  |             H
   H  H  H  H                  H  H             H
                                              /
                                           >C<
                                          H    H
            or                              or
   CH3 . CH2 . CH2 . CH3              CH3 . CH : (CH3)2.²
```

Two butanes are predicted by theory and two are actually known. This phenomenon is termed "**isomerism**," and the two compounds are termed "**isomers**." It will be noted that the two compounds (1) possess the **same** percentage composition, (2) possess the **same** molecular weight, but (3) **differ** in structure. One of the most curious phenomena discovered by chemical investigation is that which has come to be known by the name of "**isomerism**." The following illustration will give a good idea of what is implied by the term. Acetic acid, the acid con-

[1] The dots indicate linkages between the carbon atoms.

[2] The two dots do not imply double linkage between the carbon atoms but one linkage to each of the two carbon atoms.

stituent of vinegar; lactic acid, the cause of sourness in sour milk; and grape sugar, the sweetening principle of grapes are all compounds of C, H and O which contain the same percentage of each of these elements, viz. 40 °/o C. 53·3 °/o O and 6·6 °/o of H. How then comes it that these compounds are so different from each other in all their properties, both physical and chemical? The answer is that

(1) Each compound contains a different number of atoms of these elements in one molecule of the compound, e.g.

Acetic acid	$C_2H_4O_2$	Composition is represented for each by the simple formula CH_2O.
Lactic acid	$C_3H_6O_3$	
Grape sugar	$C_6H_{12}O_6$	

(2) The arrangement of the atoms in the molecule is totally different in each case, e.g.

Acetic acid $CH_3.CO.OH$.
Lactic acid $CH_2(OH).CH_2.CO.OH$.
Grape sugar $CH_2OH.CH.OH.CH.OH.CH.OH.CH.OH$ CHO.

Thus it is seen that difference in properties is due to two causes.

The second form of butane is termed **iso-butane** or **secondary butane**, and the molecule contains one carbon atom united to one hydrogen atom, viz., $-CH=$, termed the **iso-** or **secondary** group. Similarly it can be shewn that three pentanes should be possible and three pentanes are actually known, viz.:

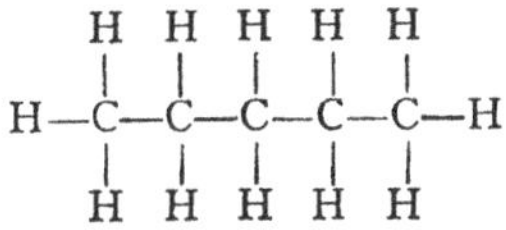

```
    H  H  H  H  H
    |  |  |  |  |
H—C—C—C—C—C—H
    |  |  |  |  |
    H  H  H  H  H
```
or $CH_3.CH_2.CH_2.CH_2.CH_3$ normal pentane

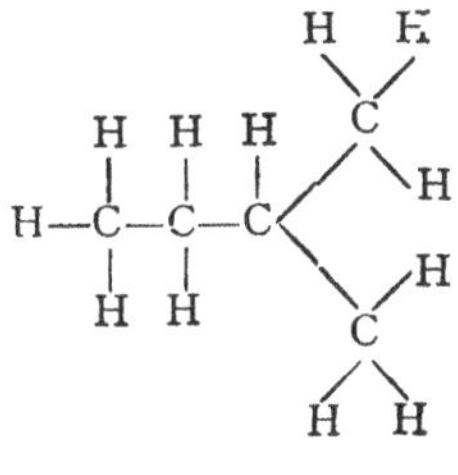

```
                H  H
                 \/
    H  H  H      C
    |  |  |     / \H
H—C—C—C<
    |  |        \ /H
    H  H         C
                 /\
                H  H
```
or $CH_3.CH_2.\mathbf{C}H:(CH_3)_2$ iso-pentane

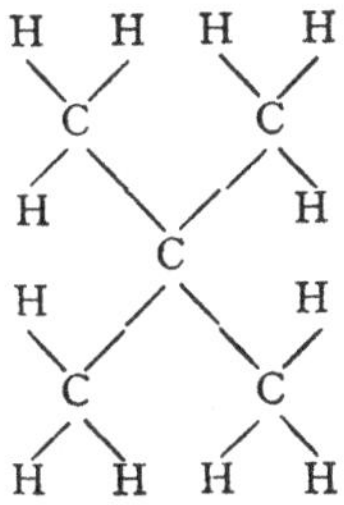

```
H   H  H   H
 \ /    \ /
  C      C
 / \    / \
H   \  /   H
     C
H   / \    H
 \ /   \  /
  C      C
 / \    / \
H   H  H   H
```
or $(CH_3)_2:\mathbf{C}:(CH_3)_2$ tertiary pentane

It will be seen that as the members of this series increase in the number of carbon atoms they contain so do the number of isomerides increase, but the latter increase at a much greater rate than the former, viz., C_6H_{14} can exist in five isomeric forms whilst C_7H_{16} can exist in nine isomeric forms, and C_8H_{18} has eighteen possible forms. These isomers can be prepared artificially, but only a few actually occur in natural products or in distillates from these. The table on page 19 gives a list of the more important paraffins.

§ 29. II. **Olefines or ethylene hydrocarbons.**

Each member of this series contains two atoms of hydrogen less than the corresponding member of the paraffin series; this is because there are two atoms of carbon in each molecule doubly linked, hence these compounds are **unsaturated** (see §§ 39 and 45).

The first few members are:

C_2H_4 or $C_2H_{(2\times 2)}$ Ethylene or Ethene
C_3H_6 or $C_3H_{(2\times 3)}$ Propylene or Propene
C_4H_8 or $C_4H_{(2\times 4)}$ Butylene or Butene
C_5H_{10} or $C_5H_{(2\times 5)}$ Pentylene or Pentene (or Amylene)

or writing a general formula which represents any member of the series

$$C_nH_{2n},$$

where n represents any whole number.

Olefines

Name	Formula	M.P.	B.P.	S.G.	Occurrence
Ethylene	C_2H_4	0° (45 at.)	−105°	—	Coal gas, shale gas
Propylene	C_3H_6	—	—	—	,, ,,
Butylenes					
Normal or α	$C_3H_6:CH_2$	—	−5°	—	,, ,,
β	$C_2H_4:C_2H_4$	—	+1°	—	—
Iso-	$(CH_3)_2:C:CH_2$	—	−6°	—	—
Amylenes					
α or normal	$C_4H_8:CH_2$	—	37°	—	—
β	$C_3H_6:C_2H_4$	—	36°	—	—
α Iso-	$(CH_3)_2CH.CH:CH_2$	—	21·3°	—	—
β Iso-	$(CH_3)_2C:C_2H_4$	—	36—38°	—	—
γ	$C_3H_8:C:CH_2$	—	—	—	—
Hexylene	C_6H_{12}	—	68°	—	Small quantities in Canadian petroleum
Heptylenes	C_7H_{14}	—	98°	—	
Octylenes	C_8H_{16}	—	124°	—	
Nonylene	C_9H_{18}	—	153°	—	
Decatylenes	$C_{10}H_{20}$	172°	—	—	—
Endecatylenes	$C_{11}H_{22}$	195°	—	—	—
Dodecylene	$C_{12}H_{24}$	−31·5°	96° (15 mm.)	·7954	—
Tetradecylene	$C_{14}H_{28}$	−12°	127°	·7936	—
Hexadecylene	$C_{16}H_{32}$	+ 4°	154°	·7917	—
Octodecylene	$C_{18}H_{36}$	18°	179°	·7910	—
Cerotene	$C_{27}H_{54}$	58°	—	—	Distillation of Chinese wax
Melene	$C_{30}H_{60}$	62°	—	—	Distillation of beeswax

The constitution of these compounds is illustrated by C_2H_4 and C_3H_6, viz.:

$$\begin{array}{cc} H & H \\ | & | \\ C & = C \\ | & | \\ H & H \end{array} \quad \text{or} \quad CH_2 : CH_2,$$

and

$$\begin{array}{ccc} H & & H \\ | & & | \\ C & = C - & C-H \\ | & | & | \\ H & H & H \end{array} \quad \text{or} \quad CH_2 : CH \,.\, CH_3.$$

The phenomenon of isomerism is also exhibited by these hydrocarbons, viz., C_4H_8 can exist in three possible forms, all of which are known, viz.:

$CH_3 \,.\, CH_2 \,.\, CH : CH_2$ normal butylene or α

$CH_3 \,.\, CH : CH \,.\, CH_3$ β butylene

$CH_2 : C : (CH_3)_2$ iso-butylene

There are five possible amylenes, etc.

Under the influence of certain compounds, e.g., dilute sulphuric acid, zinc chloride, etc., the olefines have the power of uniting with themselves to form more complex molecules, e.g., two molecules of iso-butylene combine to form one molecule of isodibutylene as represented by the following equation :

$$(CH_3)_2 : C : CH_2 + CH_2 : C : (CH_3)_2 = \underset{\text{Isodibutylene.}}{(CH_3)_3 \vdots C \,.\, CH : C : (CH_3)_2}$$

This phenomenon is known as **polymerisation**, and isodibutylene is said to be a **polymer** of isobutylene.

Similarly propylene produces polymers $(C_3H_6)_n$ and isoamylene produces the polymers di-isoamylene $(C_5H_{10})_2$, tri-isoamylene $(C_5H_{10})_3$, etc.

§ 30. III. The acetylenes.

The members of this series are characterized by the presence of a triple bond between two carbon atoms in each molecule: hence these hydrocarbons are more unsaturated than the olefines. The first few members are :

C_2H_2 or $C_2H_{(2\times2)-2}$ Acetylene or Ethine
C_3H_4 or $C_3H_{(2\times3)-2}$ Allylene or Propine
C_4H_6 or $C_4H_{(2\times4)-2}$ Crotonylene or Butine
C_5H_8 or $C_5H_{(2\times5)-2}$ Valerylene or Pentine
C_6H_{10} or $C_6H_{(2\times6)-2}$ Hexoylene or Hexine

or generally C_nH_{2n-2}.

These hydrocarbons exhibit the phenomenon of **isomerism** and **polymerism** to an even greater extent than the olefines. There are two isomeric forms of allylene, viz.:

$CH_3 \,.\, C \vdots CH$ Allylene. and $CH_2 : C : CH_2$ Isomeric allylene.

It will be noted that isomeric allylene does not possess a triple bond but two double bonds and hence it is not a true acetylene.

Crotonylene can exist in four isomeric forms, viz.:

$$CH_3 . CH_2 . C : CH, \quad CH_3 . C : C . CH_3, \quad CH_3 . CH : C : CH_2,$$
$$\text{and } CH_2 : CH . CH : CH_2,$$

i.e. in two true acetylene forms and two pseudo-forms.

Dilute sulphuric acid and high temperatures bring about polymerization, e.g.,

Acetylene at a red heat becomes partially converted to Benzene $(C_2H_2)_3$ or C_6H_6.

Allylene C_3H_4 by sulphuric acid polymerizes to Mesitylene $(C_3H_4)_3$ or C_9H_{12}, etc.

These compounds have not been found naturally in mineral oils[1] but acetylene is a product of the distillation of coal and shale and some of the ring compounds occurring naturally and in the shale distillates are probably formed by the polymerization of acetylenes.

§ 31. IV. **Naphthenes.**

The members of this series are saturated compounds, i.e., no carbon atom is linked to another carbon atom by more than one valency[2]; they differ however from paraffins in that the carbon atoms form a closed chain, known as "a ring."

The first member contains **six carbon atoms** linked to each other by one valency and all other members contain this nucleus of six carbon atoms in addition to one or more side-chains.

The structure of the first two is as shewn:

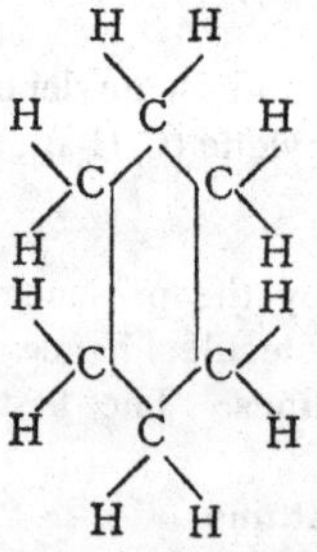

H H
C
H C C H
H H
H H
H C C H
C
H H

or $CH_2 . CH_2 . CH_2 . CH_2 . CH_2 . CH_2$ (ring closed)

or C_6H_{12}

H_2
C
H_2C $CHCH_3$
H_2C CH_2
C
H_2

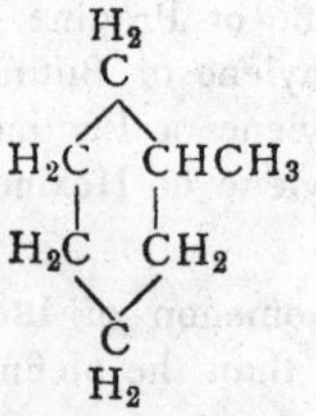

or $C_6H_{11} . CH_3$

[1] C_2H_2 is said to exist in gas from Baku wells and higher members are stated to have been isolated from Texas oils.

[2] See § 39.

The following members are known :

		M.P.	B.P.	S.G.	
Hexanaphthene	C_6H_{12}	2°	80·8°	·76 @ 0°	
Heptanaphthene	C_7H_{14}	—	101°	—	
Octanaphthene	C_8H_{16}	—	116—120°	·7714 @ 0°	Occur in Caucasian petroleum
Nonanaphthene	C_9H_{18}	—	135—140°	·7812	
Decanaphthene	$C_{10}H_{20}$	—	155—165°	·7084 $\frac{10^\circ}{4}$	
Undecanaphthene	$C_{11}H_{22}$	—	180—185°	·8019	
Dodecanaphthene	$C_{12}H_{24}$	—	196—197°	·8120	
Tetradecanaphthene	$C_{14}H_{28}$	—	240—241°	·8250	
Pentadecanaphthene	$C_{15}H_{30}$	—	246—248°	·829	

Fractions containing Naphthenes are also found in petroleums from Galicia and Burmah.

§ 32. V. Benzenes.

These hydrocarbons, like the naphthenes, are "ring" compounds or closed chain compounds. They occur to a certain extent in Pennsylvanian and Ohio petroleums as well as in the asphaltic oils of California (as much as 10 °/。 can be obtained from Caucasian petroleum); they however exist to a great extent in the crude paraffin oils obtained by the distillation of shales, and in coal tar oils. Unlike the naphthenes they are not saturated compounds nor are they unsaturated compounds like the olefines. The first member, which is the commonest, contains six atoms of carbon in the molecule forming a closed chain or ring and all the homologues are derived from this nucleus by the addition of one or more side-chains. The chief members of the series are :

		M.P.	B.P.	S.G.	Occurrence
Benzene	C_6H_6	+0° C.	80·5° C.	·8799 @ 20° C.	Coal tar oils
Toluene	C_7H_8	—	110·3°	·8656 @ 20°	Shale oils
Xylenes	C_8H_{10}	—	—	—	Pennsylvania, Ohio petroleum and asphaltic oils of California
ortho-	$C_6H_4(CH_3)_2$	(1·2)	142°	—	
meta-	$C_6H_4(CH_3)_2$	(1·3)	137°	·878 @ 0°	
para-	$C_6H_4(CH_3)_2$	(1·4)	136—137°	·862 @ 19°	
ethyl-benzene	$C_6H_5 . C_2H_5$	—	134°	·866 @ 22°	

or writing the general formula we get

$$C_nH_{2n-6},$$

where n represents any whole number.

The constitution of these compounds is illustrated by C_6H_6 and C_7H_8, viz.:

```
        H                         CH3
        |                          |
        C                          C
     /     \                    /     \
H-C           C-H          H-C           CH
  |           |              |           |
H-C           C-H          H-C           CH
     \     /                    \     /
        C                          C
        |                          |
        H                          H
```

§ **33.** It will be noted that the fourth valency of each carbon atom is not directly acting upon ony other specific atom ; as stated above there are no ethylene linkages in the molecule, i.e., no double bonds, and as there is uncertainty as to how the fourth valency is acting, the structural formula given expresses this uncertainty by assuming that the six unused valencies mutually act upon each other centrally in the molecule; this structural formula is known as the centric formula and expresses most satisfactorily the chemical reactions of the benzenes.

Isomerism in the benzene series is of two kinds, viz.:

(1) Isomerism of position of the side-chain in the nucleus ;

(2) Isomerism of the side-chain itself, as in the paraffins.

Isomerism of position.

Since the benzene ring is symmetrical only one compound results by inserting one side group, e.g., there is only one toluene which is derived from benzene by substituting a methyl group for one atom of hydrogen. When two methyl groups are substituted for two hydrogen atoms in benzene then it is possible to obtain **three** different positions of these two substituting groups, viz.:

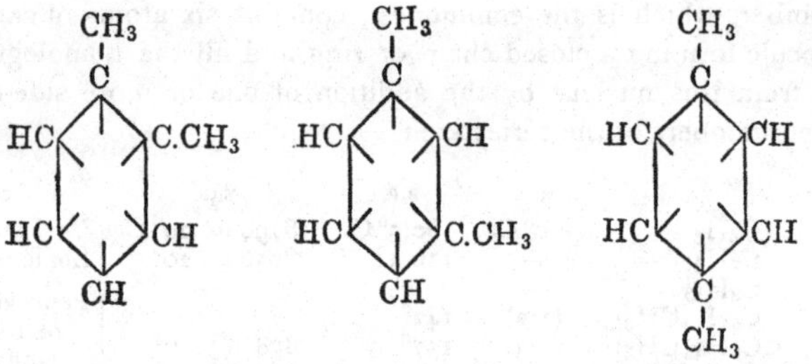

CH_3 / C / HC C.CH_3 / HC CH / CH

CH_3 / C / HC CH / HC C.CH_3 / CH

CH_3 / C / HC CH / HC CH / C / CH_3

Three such compounds are actually known. In order to differentiate between these compounds the carbon atoms in the nucleus are numbered in order from 1 to 6, viz.:

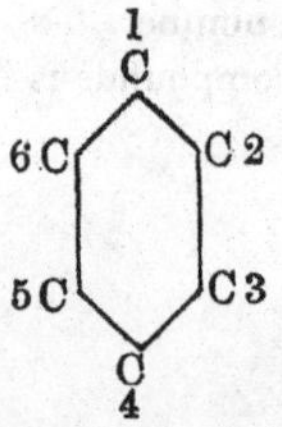

1 / C / 6C C2 / 5C C3 / C / 4

Then if two groups attach themselves to the ring at two **contiguous** carbon atoms, viz. (1 and 2, 2 and 3, 3 and 4, 4 and 5, 5 and 6, 6 and 1), the substituting groups are said to be in the **1, 2** or **ortho**-position; again if two groups enter at two carbon atoms separated by a third carbon atom, viz. (1 and 3, 2 and 4, 3 and 5, 4 and 6, 5 and 1 or 6 and 2), the substituting groups are said to be in the **1, 3** or **meta**-position, whilst if the two groups enter at two carbon atoms separated by two other carbon atoms, viz. (1 and 4, 2 and 5, 3 and 6, 4 and 1, 5 and 2, 6 and 3), the substituting groups are said to be in the **1, 4** or **para**-position.

Thus there are three dimethyl benzenes or xylenes, viz.:

1, 2 dimethyl benzene or	orthoxylene	$C_6H_4(CH_3)_2$ (1, 2)
1, 3 " "	metaxylene	$C_6H_4(CH_3)_2$ (1, 3)
1, 4 " "	paraxylene	$C_6H_4(CH_3)_2$ (1, 4)

Isomerism of the side-chain.

This isomerism is of the same nature as that of the paraffins, e.g., there are two propyl benzenes, viz.:

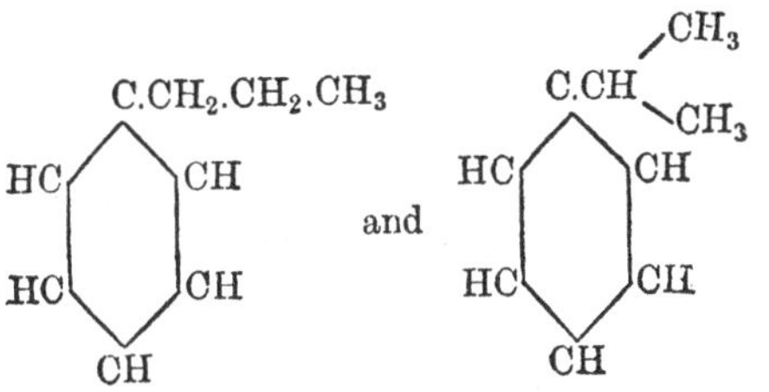

(1) benzene ring: C.CH_2.CH_2.CH_3, HC, CH, HC, CH, CH

and

(2) benzene ring: C.CH(CH_3)(CH_3), HC, CH, HC, CH, CH

No. 1 is normal propyl benzene $C_6H_5 . C_3H_7$ n.
No. 2 is iso- " " $C_6H_5 . C_3H_7$ iso.

From the above it will be seen that the possible number of isomerides increases rapidly with the number of carbon atoms, especially when it is taken into account that three or more groups or side-chains can enter the nucleus. For example the hydrocarbon $C_{10}H_{14}$ can exist in 22 isomeric forms, viz.:

$C_6H_2(CH_3)_4$	$C_6H_3 \langle^{C_2H_5}_{(CH_3)_2}$	$C_6H_4 \langle^{C_2H_5}_{C_2H_5}$	$C_6H_4 \langle^{CH_3}_{C_3H_7}$	$C_6H_5 . C_4H_9$.
3 isomerides	6 isomerides	3 isomerides	6 isomerides	4 isomerides

Some of the higher members occur in Caucasian petroleum, viz.:

(1) Pseudocumene $C_6H_3(CH_3)_3$ (1, 3, 4)

(2) Several of $C_6H_3 \langle^{CH_3}_{C_3H_7}$ and

(3) Members of $C_{11}H_{16}$.

§ **34.** Sulphur compounds. It was previously stated that some petroleums contain sulphur compounds. This is particularly so in the case of Canadian petroleum. *Mabery and Quale* have isolated and identified several such compounds belonging to a homologous series of general formula $C_nH_{2n}S$. These compounds do not contain an ethylene link, i.e., a double bond, but are probably constituted like the naphthenes. The following members have been isolated from Canadian petroleum.

		B.P. of fraction	S.G. @ 20° C.	Mol. refract.	
				Obs.	Calc. (from formula)
Hexathiophane	$C_6H_{12}S$	—	—	—	—
Heptylthiophane	$C_7H_{14}S$	158—160°	·8878	40·82	40·15
Octylthiophane	$C_8H_{16}S$	167—169°	·8929	44·71	44·71
Iso-octylthiophane	$C_8H_{16}S$	183—185°	·8937	—	—
Nonylthiophane	$C_9H_{18}S$	193—195°	·8997	49·50	49·35
Decylthiophane	$C_{10}H_{20}S$	207—209°	·9074	53·6	53·94
Undecylthiophane	$C_{11}H_{22}S$	128—130° (50 mm.)	—	58·53	58·54
Tetra-decylthiophane	$C_{14}H_{28}S$	266—268° (750 mm.)	·9208	71·61	72·33
Hexa-decylthiophane	$C_{16}H_{32}S$	283—285°	·9222	80·44	81·52
Octo-decylthiophane	$C_{18}H_{36}S$	290—295°	·9235	—	—

From the chemical behaviour of these compounds it is highly probable that each member is a ring compound of which the nucleus is either (1) a pentamethylene and sulphur ring, (2) a hexamethylene and sulphur ring, or (3) a heptamethylene and sulphur ring, viz., hexathiophane may be

```
        H2C    CH.CH3                     CH2—CH2—CH2
(1)  H2C<            >S     or  (2)  CH2<             >S
        H2C    CH2                        CH2—CH2—CH2
```

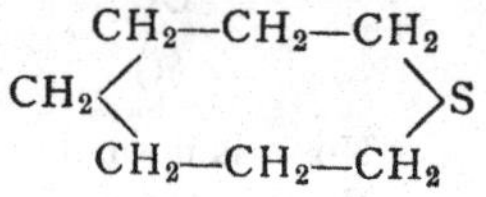

whilst heptathiophane may be

```
       CH2—CH.CH2.CH3                  CH2—CH2—CH.CH3
(1) CH2<           >S        or (2)    |             >S
       CH2—CH2                         CH2—CH2—CH2

                          CH2—CH2—CH2
              or (3)  CH2<          >S
                          CH2—CH2—CH2
```

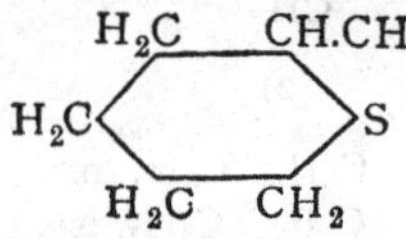

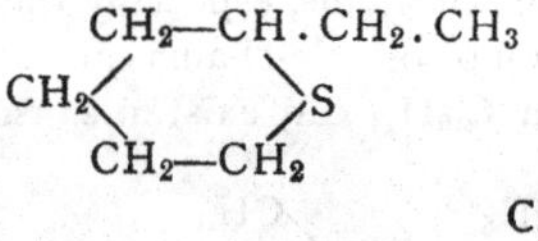

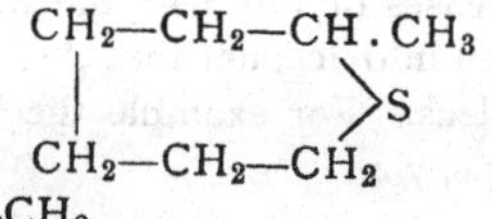

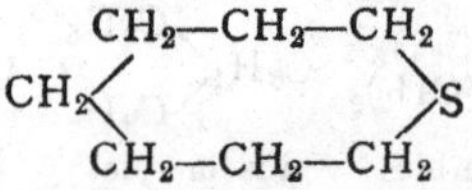

CHAPTER III

SAPONIFICATION PRODUCTS OF FATTY OILS, FATS AND NATURAL WAXES

A. **Introductory.**

§ **35.** When oils and fats are treated with steam under high pressure, either alone, or in presence of small quantities of bases, they split up, the elements of water are absorbed and glycerine is produced, together with a mixture of fatty acids. This process of **splitting** and absorption of water is termed **hydrolysis** or **saponification**. The former of these terms signifying a "water-splitting" is the more correct and will be used throughout this book. The term "saponification" will be employed only when, as the word itself implies, large quantities of bases are used, and when the aim is consequently to produce a **soap**. The reactions may be thus illustrated:

Hydrolysis, i.e. water-splitting:

$$\boxed{C_3H_5}\begin{cases} CO.OR \\ CO.OR \\ CO.OR \end{cases} + \begin{matrix} H.\boxed{OH} \\ H.\boxed{OH} \\ H.\boxed{OH} \end{matrix} = C_3H_5\begin{cases} OH \\ OH \\ OH \end{cases} + \begin{matrix} R.CO.OH \\ R.CO.OH \\ R.CO.OH \end{matrix}$$

1 molecule of glyceride — 3 molecules of water — 1 molecule of glycerine — 3 molecules of fatty acid

Saponification, i.e. soap-making:

$$\boxed{C_3H_5}\begin{cases} CO.OR \\ CO.OR \\ CO.OR \end{cases} + \begin{matrix} Na.\boxed{OH} \\ Na.\boxed{OH} \\ Na.\boxed{OH} \end{matrix} = C_3H_5\begin{cases} OH \\ OH \\ OH \end{cases} + \begin{matrix} R.CO.ONa \\ R.CO.ONa \\ R.CO.ONa \end{matrix}$$

1 molecule of glyceride — 3 molecules of sodium hydrate (a base) — 1 molecule of glycerine — 3 molecules of a sodium salt, i.e. a soap

§ **36.** The word **soap**, although applied commonly only to the alkali salts of the fatty acids is, strictly speaking, the name of all combinations of a mineral base with a fatty acid. Potash and soda soaps are respectively the "soft" and "hard" soaps of commerce and are the potassium and sodium salts of **any** fatty acid. Ammonium soaps also find some uses. These three kinds of soap are all soluble in hot water. The soaps formed by the treatment of oils with lime, calcium oxide (CaO); magnesia, magnesium oxide (MgO); zinc oxide (ZnO); or by the combination of a fatty acid with any inorganic base other than the soluble alkalies are termed the "metallic soaps." They are insoluble in water, and several of them are of considerable commercial importance.

A soap is a true **salt**. Just as on treatment of a mineral acid with a mineral base there results the formation of a salt and water, so when a **fatty** acid reacts with a mineral base, a **soap** or salt of the fatty acid is formed : e.g. compare the following chemical reactions.

$$\boxed{H}Cl + Na\boxed{OH} = NaCl + H.OH$$

mineral acid | mineral base | salt | water

and

$$R.CO.O\boxed{H} + Na\boxed{OH} = R.CO.ONa + H.OH$$

fatty acid | mineral base | salt or soap | water

§ **37. Hydrolysis**, or fat splitting, is produced commercially by the following methods :

1. Treatment with high pressure steam assisted by small quantities of bases.
2. Treatment with strong mineral acid (sulphuric) followed by steaming.
3. Boiling with water containing a stable fatty acid compound of sulphuric acid (the "saponifier").
4. Treatment with water and a certain amount of a substance known as a "ferment" or "enzyme" extracted from castor seed.

All these processes result in producing a large proportion of fatty acids and glycerine from the oil or fat, leaving varying amounts of the original oil unsplit.

B. The Fatty Acids.

§ **38.** Chemically the fatty acids are compounds of carbon, hydrogen and oxygen; they are all similar to the extent that the molecule of every fatty acid contains a characteristic group of atoms situated at the end of the chain of carbon atoms, known as the **carboxyl group** in which the atoms are attached to one another as shewn in the following structural formula :

$$-C\begin{matrix}\nearrow O \\ \searrow O-H\end{matrix}$$

or written in the contracted form

$$.CO.OH.$$

The simplest member of the series, namely Formic Acid, contains another H atom attached to the free valency of the carbon atom of the carboxyl group and hence its structure is H.CO.OH. Formic acid however is not a true fatty acid inasmuch as it does not occur in oils or fats.

§ **39.** The fatty acids are classified into two important sections, viz. **saturated** compounds and **unsaturated** compounds. In order that the reader may grasp the full meaning of this most important distinction it will be helpful to recall some of the fundamental ideas of chemical theory and shew their bearing on the significance of these terms.

All the simple chemical substances, the elements, consist of an agglomeration of indivisible[1] particles or **atoms** (an atom may be defined as the smallest part of an element which is capable of exhibiting the characteristic chemical properties of that element). All atoms possess certain definite properties, e.g.

(i) The atoms of the same element possess a definite and constant weight.

(ii) The atoms exhibit certain **preferences**, under certain conditions, to combination with other atoms, termed chemical affinities.

(iii) The atoms have certain definite powers of combination called **valencies.**

It is with this last property that we here have to deal. It is found from experience that **one atom** of hydrogen[2] never combines with more than **one atom** of any other element and hence the atom of hydrogen is said to be mono-valent, i.e. has **one power** of combination. The atom of oxygen, however, is able to combine with **two atoms** of hydrogen and hence it is termed a **di-valent** element. There are several other elements, like oxygen, that are also di-valent, e.g. copper, zinc, etc. ; **one atom** of copper combines with only **one atom** of oxygen forming a molecule of copper oxide, CuO.

One atom of the element nitrogen is capable of combining with **three atoms** of hydrogen and hence nitrogen is said to be **trivalent.** One atom of carbon unites with **four atoms** of hydrogen and is thus a **tetravalent** element ; similarly there are penta-, hexa-, hepta- and octa-valent elements.

In order to obtain a better mental picture of the mechanism of chemical reactions, it has become the custom to represent the valency of an element graphically by drawing a line, attached to the symbol of the element, for each valency, possessed by the atom, e.g.

$$\mathrm{H}-,\ -\mathrm{O}-,\ -\overset{|}{\mathrm{N}}-,\ -\overset{|}{\underset{|}{\mathrm{C}}}-,\ \text{etc.}$$

[1] Recent investigations shew that the atom is divisible but this does not invalidate the Atomic Theory, since no chemical reaction is known in which the atom does not act as a whole.

[2] There are certain apparent exceptions to this.

In a work of this character, written not for chemical students only, but for practical men, it is necessary to confine the discussion of chemical theory to the simplest possible statements and hence no evidence is adduced for the same. There are many exceptions to some of the statements made but they do not invalidate the main argument.

The lines represent the force or forces exerted by the atom in combining with other atoms, irrespective of magnitude or direction[1].

It has already been stated that one of the characteristic properties of the carbon atom is its power of combining with atoms of itself. The following simple and well-known compounds of carbon illustrate the modes of combination of carbon atoms:

(1) Ethane, a gas occurring to a small extent in coal gas and having the composition C_2H_6.

(2) Ethylene, another gas occurring to a much larger extent and represented by the formula C_2H_4.

(3) Acetylene, the well-known illuminant, whose formula is C_2H_2.

Now it has been shewn that the various valencies in these compounds are distributed between the atoms as shewn in the following **structural formulae**:

$$\begin{array}{ccc} \begin{array}{c} \text{H}\ \ \text{H} \\ |\ \ \ | \\ \text{H—C—C—H} \\ |\ \ \ | \\ \text{H}\ \ \text{H} \end{array} & \begin{array}{c} \text{H}\ \ \text{H} \\ |\ \ \ | \\ \text{C—C} \\ |\ \ \ | \\ \text{H}\ \ \text{H} \end{array} & \text{H—C—C—H.} \\ (1) & (2) & (3) \end{array}$$

It will be observed that in all the formulae the atoms of H are shewn to exert one **force** or valency, whilst in (1) the carbon atoms each exert **4 valencies**, in (2) each carbon atom exerts **3 valencies**, and in (3) each carbon atom exerts only **2 valencies**.

All compounds in which the carbon atoms as in (1) exert a valency of **four** are said to be **saturated**, whilst compounds like (2) and (3) where the valencies of the carbon atoms appear to be 3 and 2 respectively are said to be **unsaturated**.

Again it is supposed that, in compounds containing two or more carbon atoms, these do each exert four valencies and that if the latter are not used up in combining with atoms of other elements they are in some way[2] used to combine the carbon atoms themselves; thus in ethylene and acetylene the valencies are supposed to be used as shewn in the following formulae:

$$\begin{array}{cc} \begin{array}{c} \text{H}\ \ \text{H} \\ |\ \ \ | \\ \text{C}=\text{C} \\ |\ \ \ | \\ \text{H}\ \ \text{H} \end{array} & \text{H—C}\equiv\text{C—H,} \end{array}$$

i.e. in ethylene the two carbon atoms act on each other with two valencies, spoken of as a **double linking** (or bond), whilst in acetylene the two carbon atoms act upon each other with three valencies called a **triple linking**.

[1] The question of the spatial arrangement of the carbon valencies will be discussed later.

[2] The manner of combination will be indicated later.

All carbon compounds containing double and triple linkings are said to be **unsaturated**.

The **unsaturated** fatty acids contain **one** or more **double linkings** in the molecule.

The photographs Figs. 1, 2 and 3 represent models of carbon compounds. The large balls represent carbon atoms and the small balls different atoms or groups of atoms combined with the carbon atoms. The models in figs. 2 and 3 show how two carbon atoms unite by two or three valencies. It will be seen that the forces are bent out of a straight

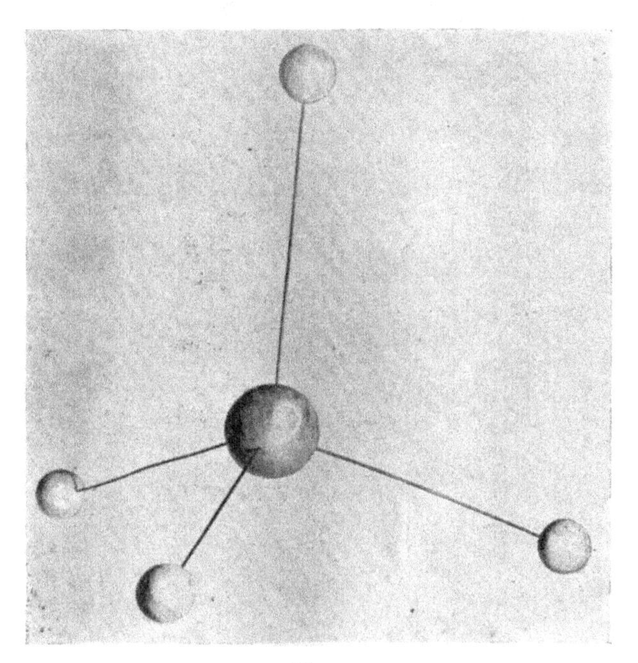

Fig. 1

line and therefore a strain is set up in the molecule, just as in the ribs of an umbrella when it is spread. Consequently the union between two carbon atoms by two or three forces or valencies is unstable and slight chemical forces are able to break them apart; hence such compounds are easily decomposed or easily enter into chemical reaction. (Baeyer's *Strain Theory*[1]).

§ **40.** There are five principal series of fatty acids.

1. The **Stearic** acid series; all saturated acids.
2. The **Oleic** acid series; all containing one double linking.

[1] See page 80.

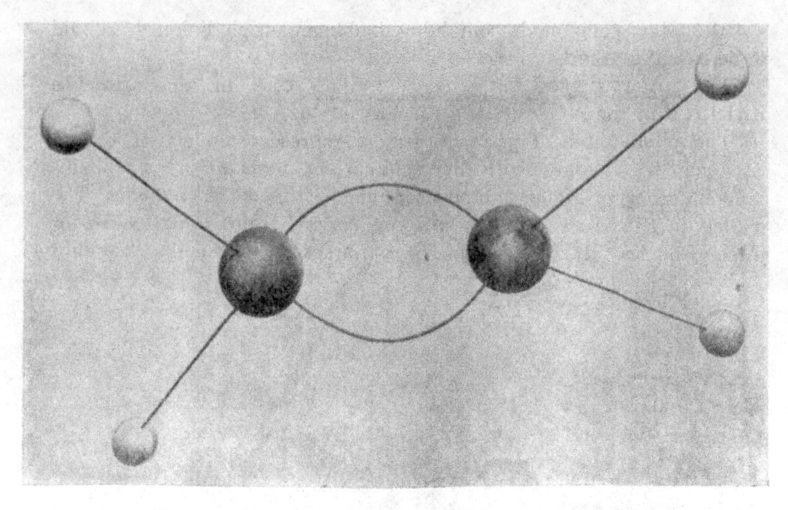

Fig. 2

Fig. 3

3. The **Linolic** acid series; containing two double linkings in the molecule.
4. The **Linolenic** acid series; containing probably three double linkings.
5. The **Clupanodonic** acid series; containing probably four double linkings.

The saturated fatty acids are distinguished practically from the unsaturated fatty acids by the fact that they are unable to **directly combine** with bromine whilst the latter can, one molecule of an unsaturated fatty acid being able to **combine directly** with 2, 4, 6 or 8 atoms of bromine according to whether it contains 1, 2, 3 or 4 double linkings.

§ 41. The Stearic Acid Series.

The simplest member of this series, as previously stated, is Formic Acid. The members of the series increase in complexity by regular differences of one atom of C and two of hydrogen, i.e. by CH_2. The structural formulae of the first three members being:

Formic acid $\quad H{-}C{\overset{\displaystyle /\!\!/O}{\underset{\displaystyle \backslash O{-}H}{}}}$ $\quad$ or $\quad H.CO.OH$

Acetic " $\quad H{-}\underset{H}{\overset{H}{C}}{-}C{\overset{\displaystyle /\!\!/O}{\underset{\displaystyle \backslash O{-}H}{}}}$ $\quad$ or $\quad CH_3.CO.OH$

Propionic " $\quad H{-}\underset{H}{\overset{H}{C}}{-}\underset{H}{\overset{H}{C}}{-}C{\overset{\displaystyle /\!\!/O}{\underset{\displaystyle \backslash O{-}H}{}}}$ $\quad$ or $\quad CH_3.CH_2.CO.OH$

A very curious and important point will be noticed regarding fatty acids which occur in fats, oils and waxes. Only **two** (valeric acid and daturic) contain an **odd number of carbon atoms** in the molecule, all the others having an **even** number[1].

As formic and acetic acids do not occur in oils and fats they are omitted in the following table of fatty acids:

[1] Acids having an odd number of carbon atoms, which were in the past believed to occur in certain fats, have been shewn to be mixtures of two or more acids of even numbers of carbon atoms. It is even possible that the Valeric acid, which is stated to occur in Porpoise and Dolphin oils, is a mixture of Butyric and Caproic acids. It is known that certain mixtures of fatty acids behave as regards melting and solidifying points like pure chemical substances. Thus a mixture of 45·5 % Stearic acid and 52·5 % Palmitic acid cannot be separated by crystallization from alcohol; it is a "eutectic mixture."

Stearic Acid Series $C_nH_{2n+1}CO.OH$

Name	Formula	No. of C atoms	S.G. water @ 4° C. = 1	M.P. ° C.	Solubility	Occurrence
Butyric acid	$CH_3.(CH_2)_2.CO.OH$	4	·9580 @ 14°	− 6·5	Miscible with water	Cow butter fat
[Valeric ,,	$CH_3.(CH_2)_3.CO.OH$	5	·9415 @ 20°	− 18 (S.P.)	4 % in water @ 15° C.	Porpoise and dolphin oils?]
Caproic ,,	$CH_3.(CH_2)_4.CO.OH$	6	·924 ,,	− 8 (S.P.)	·9 % ,, ,,	Cocoanut oil group
Caprylic ,,	$CH_3.(CH_2)_6.CO.OH$	8	·9100 ,,	16·5	·08 % ,, ,,	Cocoanut oil group
Capric ,,	$CH_3.(CH_2)_8.CO.OH$	10	·8858 @ 40°	31·3	1 in 1000 in boiling water	Cocoanut oil group
Lauric ,,	$CH_3.(CH_2)_{10}.CO.OH$	12	·875 @ 43·6°	43·6	Slightly sol. in boiling water	Cocoanut, laurel, etc.
Myristic ,,	$CH_3.(CH_2)_{12}.CO.OH$	14	·8622 @ 53·8°	53·8	Insol. in water; diff. sol. in cold alcohol	Myristica group; laurel, etc.
Palmitic ,,	$CH_3.(CH_2)_{14}.CO.OH$	16	·8527 @ 62°	62·6	6 % in alcohol @ 15° C.	Most fats
Daturic[1] ,,	$CH_3.(CH_2)_{15}.CO.OH$	17	·8532 @ 60°	59·5	Insol. in water, becoming decreasingly soluble in alcohol	Datura oil
Stearic ,,	$CH_3.(CH_2)_{16}.CO.OH$	18	·8454 @ 69·2°	69·3	Insol. in water, becoming decreasingly soluble in alcohol	Most fats
Arachidic ,,	$CH_3.(CH_2)_{18}.CO.OH$	20	—	77	Insol. in water, becoming decreasingly soluble in alcohol	Arachis oil
Behenic ,,	$CH_3.(CH_2)_{20}.CO.OH$	22	—	80—82	Insol. in water, becoming decreasingly soluble in alcohol	Oil of ben
Lignoceric ,,	$CH_3.(CH_2)_{22}.CO.OH$	24	—	80·5	Insol. in water, becoming decreasingly soluble in alcohol	Arachis oil
Cerotic ,,	$CH_3.(CH_2)_{24}.CO.OH$	26	·8359 @ 79°	77·8	Soluble in boiling alcohol	Beeswax and montan wax
Montanic ,,	$CH_3.(CH_2)_{26}.CO.OH$	28	—	83	Slightly sol. in methyl alcohol	Montan wax
Melissic ,,	$CH_3.(CH_2)_{28}.CO.OH$	30	—	91	Insoluble in methyl alcohol	Beeswax and montan wax

[1] Meyer and Beer, "Datura oil," *Käiserl. Akad. d. Wissenschaften*, Wien, Jan. 1912.

§ **42**. It will be noticed that with the **increase** in the number of carbon atoms (and consequent increase in the molecular **weight**),

(*a*) The specific gravity decreases.
(*b*) The solubility decreases.
(*c*) The melting point increases.

The lower members of the series are liquids; the higher members are hard white crystalline solids, whilst the intermediate members are buttery in consistence. Besides being more or less soluble, as stated, in water the lower fatty acids, butyric, valeric, caproic, caprylic, and capric are volatile in steam, i.e. on boiling with water these acids tend to pass over with the steam and can hence be condensed with the steam. This fact is made use of in the Reichert test applied to those fats which contain **lower** or **volatile** fatty acids, e.g. butter, cocoanut, etc.

§ **43. Acetic acid.** $CH_3 . CO . OH$.

This acid has been stated to occur in the seeds of the spindle tree. Confirmation is necessary. For properties of this acid see text-books on organic chemistry.

Butyric acid. $CH_3 . (CH_2)_2 . CO . OH$.

This acid occurs as a mixed glyceride in butter-fat to the extent of 6°/ₒ. Colourless liquid crystallizing at $-19°$ C.; its aqueous solution recalls the odour of rancid butter. On warming an alcoholic solution of the acid with sulphuric acid, ethyl butyrate $CH_3(CH_2)_2CO . OC_2H_5$ is formed (liquid with pine-apple odour).

Valeric acid. $CH_3 . (CH_2)_3 . CO . OH$.

It is doubtful whether this acid occurs naturally or not. It is stated to occur in the two blubber oils from the porpoise and dolphin, but it is probable that the acid taken to be valeric is a mixture of butyric and caproic acids.

Odour resembles putrid cheese.

Colourless liquid. B.P. 186° C. Solidifying at $-18°$ C. Soluble at 16° C in 27 volumes of water.

Caproic acid. $CH_3 . (CH_2)_4 . CO . OH$.

Occurs together with butyric acid and the two following acids in butter-fat, cocoanut and palm nut oils. Has an odour resembling that of sweat. It is distinguished from butyric acid by the fact that zinc caproate $\{CH_3 . (CH_2)_4 . CO . O\}_2Zn . H_2O$ is precipitated on adding the acid to a solution of zinc acetate.

100 parts of water dissolve 11·1 parts of Barium caproate @ 10·5° C.

Caprylic acid. $CH_3 . (CH_2)_6 . CO . OH$.

Resembles the preceding acid. It is a liquid which crystallizes on cooling to 12° C. in plates melting at 16·5° C.

100 parts of water dissolve 0·619 parts of Barium caprylate at 20° C.

The Ethyl ester $CH_3 . (CH_2)_6 . CO . OC_2H_5$ boils at 207–208° C.

Capric acid. $CH_3.(CH_2)_8.CO.OH$.

Crystalline solid. M.P. 31·3° C.

Barium caprate nearly insoluble in cold water and sparingly soluble in boiling water.

The Ethyl ester boils at 243–245° C.

Lauric acid. $CH_3.(CH_2)_{10}.CO.OH$.

Occurs in cocoanut oil, palm nut oil, laurel oil and in the less known tangkallak and dika fats.

Crystallizes from alcohol in needles. M.P. 43·6° C.

The previously mentioned acids can all be distilled at atmospheric pressure unchanged. Lauric acid, however, and all the acids of higher molecular weight in this series undergo decomposition unless distilled under reduced pressure.

100 parts of boiling water dissolve only 0·07 parts of Barium laurate.

The Ethyl ester solidifies at – 10° C.

Myristic acid. $CH_3.(CH_2)_{12}.CO.OH$.

This acid forms a large proportion of the lesser known fats belonging to the "Myristica[1]" group. To a less degree it occurs in cocoanut oil, palm nut oil, lard, linseed oil, arachis oil, butter-fat and several other fats. It probably also occurs combined with monohydric alcohols in spermaceti and wool wax.

Completely insoluble in water.

M.P. 53·8° C.

Barium myristate is almost insoluble in boiling water.

The Ethyl ester solidifies at 10·5–11·5° C.

Palmitic acid. $CH_3.(CH_2)_{14}.CO.OH$.

This acid occurs in most vegetable and animal fats. In large quantities in Palm oil, Chinese vegetable tallow, Japan-wax and Myrtle-wax (Myrica-tallow),

also in Spermaceti as cetyl palmitate
Beeswax „ myricyl „
Opium-wax „ ceryl „

Palmitic acid can be produced from oleic acid by melting the latter with caustic potash (KOH); the reaction is as follows:

$$\underset{\text{oleic acid}}{C_{17}H_{33}.CO.OH} + \underset{\text{caustic potash}}{2\,KOH} = \underset{\text{potassium palmitate}}{C_{15}H_{31}.CO.OK} + \underset{\text{potassium acetate}}{CH_3.CO.OK} + \underset{\text{hydrogen}}{H_2}$$

Oxalic acid is stated to be formed during the reaction, which once formed the basis of a technical method (Varrentrap's process).

Slightly soluble in cold alcohol (about 9 °/₀ at 19° C.). Readily soluble in boiling alcohol.

Barium palmitate dissolves to the extent of ·0035 parts in 100 parts of absolute alcohol.

[1] Gr. *murizo*=to be fragrant with ointment, referring to the odour of the fruit. Myristacea is the name of a botanical order to which the nutmeg belongs.

Its Ethyl ester has M.P. 24·2° C.

The acid possesses M.P. 62° C.

Daturic acid[1]. $CH_3.(CH_2)_{15}.CO.OH$.

This acid has recently been proved to exist in Datura oil. It forms the notable exception of being the only acid with an uneven number of carbon atoms proved to exist in an oil.

M.P. 59·5° C.; more soluble in alcohol than Palmitic acid; Ethyl ester has M.P. 26·7° C.

Stearic acid. $CH_3.(CH_2)_{16}.CO.OH$.

This acid is very abundant in all fats and many oils. It is also obtained by the catalytic hydrogenation of the unsaturated fatty acids, containing the same number of carbon atoms, e.g. oleic, linolic, linolenic and clupanodonic.

When boiled under ordinary pressure it distils at 360° C. but suffers slight decomposition; under a partial vacuum it can be distilled unchanged.

It is insoluble in water.

Hot alcohol dissolves it readily; 100 parts of cold absolute alcohol dissolve about 2·5 parts of stearic acid.

Barium stearate is practically insoluble in alcohol.

The Ethyl ester has M.P. 36·7° C.

The acid has M.P. 69·32° C.

Arachidic acid. $CH_3.(CH_2)_{18}.CO.OH$.

Occurs naturally in Arachis oil and in smaller quantities in rape oil, cacao butter and cow butter.

Sparingly soluble in cold alcohol; dissolves readily in boiling alcohol.

Ethyl ester. M.P. 50° C.

Crystals of acid. M.P. 77° C.

Behenic acid. $CH_3.(CH_2)_{20}.CO.OH$.

Occurs in Ben oil.

100 parts of alcohol dissolve 0·102 parts of the acid at 17° C.

Ethyl ester. M.P. 48° C. (Voelcker).

M.P. 80–82° C. The synthetic acid prepared by the hydrogenation of Erucic acid has M.P. 83–84° C.

Lignoceric acid. $CH_3.(CH_2)_{22}.CO.OH$.

Occurs in Arachis oil.

Ethyl ester. M.P. 55° C.

Acid. M.P. 80·5° C. (?).

Cerotic acid. $CH_3.(CH_2)_{24}.CO.OH$.

Occurs free in beeswax, montan wax and in carnaüba wax; also in insect wax and wool wax as ceryl cerotate.

The crude acid is prepared from beeswax by exhaustion with boiling alcohol.

M.P. 77·8° C.; soluble in methyl alcohol.

Ethyl ester. M.P. 59–60° C.

[1] Meyer and Beer, *Kaiserl. Akad. d. Wissenschaften*, Wien, Jan. 1912.

Montanic acid. $CH_3.(CH_2)_{26}.CO.OH$[1].

Occurs free in "Montan Wax," the distilled product of brown coal bitumen.

Crystalline needles. M.P. 83° C.

Soluble in hot alcohol; slightly soluble in methyl alcohol.

A diglyceride has been formed $C_3H_5(C_{28}H_{57}.CO.O)_2OH$.

M.P. 83° C.; ethyl ester M.P. 67° C.

Melissic acid. $CH_3.(CH_2)_{28}.CO.OH$.

Occurs free in beeswax and in montan wax.

Sparingly soluble in ether, soluble in hot alcohol, insoluble in cold methyl alcohol.

Ethyl ester. M.P. 73° C.

M.P. 91° C.

§ **44. The Oleic acid series.** $C_nH_{2n-1}CO.OH$.

The acids belonging to this series are unsaturated compounds. They contain **one double linking** between the carbon atoms in the molecule; they are therefore capable of combining directly with two atoms of hydrogen being converted into a saturated acid of the stearic acid series. This hydrogenation only takes place however in the presence of a catalyst (e.g. finely divided nickel or palladium).

The acids of this series are more soluble in alcohol than the corresponding saturated fatty acids. The lead salts of this series, together with those of the less saturated acids, are soluble in ether; a property which is made use of in separating them from the saturated acids whose lead salts are insoluble in ether.

When treated with nitrous acid they undergo a molecular rearrangement producing compounds known as "Elaidins" (see note on Stereoisomerism, § 45).

On oxidation with a dilute alkaline solution of potassium permanganate they yield hydroxy acids, in which reaction one of the two double bonds between two carbon atoms is ruptured and union with – OH radicles takes place, viz. —CH = CH— becomes

$$\begin{array}{cc} —CH— & CH— \\ | & | \\ OH & OH \end{array}$$

According to the position of the doubly linked carbon atoms various isomers are possible of which a few have been isolated from oils and fats.

The methods of determining the position of the double linking are as follows, the example given being Oleic acid:

1. By conversion of the acids into "stearolic" acids.

(*a*) The acids are treated with Br which directly combines with the acid to form a dibromide, viz.:

$$CH_3.(CH_2)_7CH=CH.(CH_2)_7CO.OH+Br_2$$
$$=CH_3.(CH_2)_7CHBr—CHBr.(CH_2)_7.CO.OH.$$

[1] Ryan and Dillon, *Sci. Proc. Roy. Dub. Soc.* 1909 (12), 202.

(*b*) The dibromide is heated with an alcoholic solution of caustic potash under pressure when a stearolic acid is formed having a treble linking.

$$CH_3 . (CH_2)_7 . C\,HBr—C\,HBr . (CH_2)_7 . CO . OH + 3KOH$$
$$= CH_3 . (CH_2)_7 . C \equiv C . (CH_2)_7 CO . OK + 2KBr + 3H_2O.$$

(*c*) Treatment with concentrated sulphuric acid converts the stearolic acid into a Keto-acid :

$$CH_3 . (CH_2)_7C \equiv C . (CH_2)_7 CO . OH + H_2O \text{ (contained in the sulphuric acid)}$$
$$= CH_3 . (CH_2)_7 \underset{\underset{O}{\|}}{C}—CH_2 . (CH_2)_7 . CO . OH.$$

(*d*) The Keto-acid gives with hydroxylamine hydrochloride two stereo-isomeric[1] oximes, viz. :

$$2CH_3 . (CH_2)_7 . \underset{\underset{O}{\|}}{C}—CH_2 . (CH_2)_7 . CO . OH + 2NH_2 . OH$$
$$= CH_3 . (CH_2)_7 . \underset{\underset{N . OH}{\|}}{C}—CH_2 . (CH_2)_7 . CO . OH \ (1) \text{ and}$$
$$CH_3 . (CH_2)_7 . \underset{\underset{HO . N}{\|}}{C}—CH_2 . (CH_2)_7 . CO . OH \ (2) + 2H_2O.$$

(*e*) With concentrated sulphuric acid these oximes yield two amino-acids, viz. :

$$CH_3 . (CH_2)_7 . \underset{\underset{O}{\|}}{\overset{10}{C}} . (NH_2) \underset{9}{CH} (CH_2)_7 . CO . OH \text{ (A) and}$$

$$CH_3 . (CH_2)_6 . CH (NH_2) . \underset{\underset{O}{\|}}{\overset{10}{C}} . \underset{9}{CH_2} (CH_2)_7 . CO . OH . \text{ (B).}$$

(*f*) These amino-acids when treated with concentrated hydrochloric acid yield the two acids :

$$CH_3 . (CH_2)_7 . \overset{10}{C}O . OH. \quad \text{Nonylic acid.}$$

$$\overset{9}{C}O . OH . (CH_2)_7 . CO . OH. \quad \text{Azelaic acid.}$$

Hence it is concluded that the double link in oleic acid occurs between the 9th and 10th carbon atom.

2. By oxidation with ozone (O_3). Ozone is passed through a solution of the acid in hexane with the production of "ozonides." On boiling these with water they are split up into :

(1) Nonylaldehyde $C_8H_{17} . CHO$ and
(2) Azelaic acid semi-aldehyde $CO . OH . (CH_2)_7 . CHO$.

3. Conversion of nitrogen peroxide addition products :

This method yields information of the same character as the previous methods. The products split into two derivatives (as above) on treatment with hydrochloric acid in sealed tubes.

[1] See § 45.

Acids of the Oleic Acid Series.

Acids up to C_4 not known in oils and fats.

Tiglic acid[1], an acid occurring in croton oil of constitution

$CH_3 . CH : C(CH_3) . CO . OH.$ M.P. 64·5° C.

Acids C_6—C_{11} not known in oils.

Acids C_{12} and C_{14} are said to occur in the fat of cochineal[2].

Acids C_{16}. Four acids with this composition are said to exist in oils and fats, viz.:

(i) **Hypogœic acid**[3] in Arachis oil: probable constitution is

$CH_3 . (CH_2)_7 . CH : CH . (CH_2)_5 . CO . OH.$ M.P. 33–34° C.

(ii) **Physetoleic acid**[4] in Caspian seal oil. M.P. 30° C. (?).

(iii) **Palmitoleic acid**[5] in cod-liver oil. S.P. − 1·5° C.

(iv) **Lycopodic acid**[6] in spores of lycopodium.

Acids C_{18}. Five of these have been found in oils and fats, but a much greater number can exist, according to the position of the double bond. Fokin[7] points out that probably the oleic acids in which the doubly linked carbon atoms are in an odd-even position are solid, whilst those in which it is even-odd are liquid.

(i) **Oleic acid** (ordinary), double link between 9 and 10 carbon atoms[8] (for the method of determination of the position of the double bond see above), viz.:

$$\begin{array}{l} CH_3 . (CH_2)_7 . \overset{10}{C}H \\ \qquad\qquad\qquad \| \\ \qquad\qquad\qquad \underset{9}{C}H(CH_2)_7 . CO . OH \end{array}$$

Olein is found in most oils and in many fats. It is obtained commercially as a bye-product in candle manufacture. It is difficult to obtain in a pure state, but when so, it is a colourless liquid without odour.

Properties Solidifies at 4° C. to needles which melt at 14° C. S.G. at 15° C. is ·898. Ref. Index $\frac{20}{D}$ = 1·4620.

It **distils** unchanged at 250° C. with superheated steam. Insoluble in water, soluble in dilute and strong alcohol.

Reactions. On treatment with nitrous acid it produces a solid isomer known as "Elaidic acid."[9]

On **exposure to the air and light** oleic acid turns yellow and becomes "rancid" with the formation of lower fatty acids which redden blue litmus paper. On **blowing air** through the acid, its specific gravity rises and "oxidized" acids (insoluble in petroleum ether) are formed.

[1] *Liebigs Ann.* 283, 65.
[2] Raymann, *Monatsch. f. Chem.* 1885 (6), 895.
[3] *Liebigs Ann.* 1855 (94), 230.
[4] Ljvbarsky, *Jour. f. Prakt. Chem.* 1898 (57), 26.
[5] *Berichte*, 1906, 3574.
[6] Langer, *Arch. d. Pharm.* 1889 (27), 241.
[7] *Jour. Russ. Phys. Chem. Soc.* 1912 (44), 653.
[8] Molinari and Soncini, *Ber.* 1906 (39), 2735.
[9] See § 45.

Sulphur when digested with oleic acid at 150° C. is absorbed without decomposition of the acid ; at higher temperature, however, sulphuretted hydrogen, H_2S, is evolved.

On oxidation with **nitric acid**, dibasic acids are produced, such as suberic acid $(CH_2)_6 : (CO . OH)_2$ and pimelic acid $(CH_2)_5 : (CO . OH)_2$, in addition to volatile acids.

On oxidation with dilute **alkaline $KMnO_4$** solution, **dihydroxystearic acid** is the chief product (see § 51). When treated with **concentrated sulphuric acid** in the cold, oleic acid combines to form a substance of probable formula

$$C_{17}H_{34}\begin{matrix} \diagup O . SO_2 . OH \\ \diagdown CO . OH \end{matrix}$$ [1] Hydroxystearo-sulphuric acid ;

on boiling with water, a mixture of hydroxystearic acid and its anhydrides is formed. The reaction will be more fully discussed in a later volume. Oleic acid absorbs two atoms of **bromine** per molecule, forming a saturated compound

$$CH_3 . (CH_2)_7 . CHBr . CHBr . (CH_2)_7 . CO . OH.$$

Hydriodic acid and phosphorus reduces the acid and subsequent treatment with zinc and hydrochloric acid (nascent H) adds on two atoms of H with the production of stearic acid.

The **reduction** of oleic acid by treatment with hydrogen in the presence of a metallic catalyst, is now of commercial importance. The history of the attempts to reduce the acid are discussed under Hardening of oils (chapter XX).

The following table gives the solubility of the chief oleates in various solvents.

Name of salt	1 part of salt soluble in			Remarks
	Water	Alcohol	Ether	
Sodium Oleate	10 parts	20 parts	100 parts (boiling)	—
Potassium ,,	4 ,,	2 ,,	30 parts	—
Calcium ,,	Insol.	Insol.	Insol.	Sl. sol. in conc. sugar solution
Barium ,,	,,	Sparingly sol. (boiling)	?	Sparingly sol. in hot benzene
Aluminium ,,	,,	Insol.	Sparingly sol.	—
Silver ,,	,,	,,	almost insol.	—
Copper ,,	,,	Sol. to green soln.	Readily sol. to green soln.	Melts at 100° to green oil
Lead ,,	,,	Sl. sol.	Sol.	Melts at 80° to yellow oil

[1] Saytzoff, *Jour. f. Prakt. Chem.* 35, 369, 57 (1898), 26.

(ii) **Iso-oleic acid** probably double link (10, 11), viz.:

$$\begin{array}{l} CH_3.(CH_2)_6.\overset{11}{C}H \\ \qquad\qquad\quad \| \\ \qquad\qquad\quad \underset{10}{C}H.(CH_2)_8.CO.OH \end{array} \qquad \text{M.P. } 24^\circ \text{C.}$$

This acid is formed on distilling hydroxystearic acid, as in production of candle material by acid saponification and distillation process. The so-called "iso-oleic" acid has been shewn to be a mixture of acids[1], including ordinary 9, 10 oleic acid and elaidic acid, an 8, 9 elaidic acid and a hydroxystearic acid. Iso-oleic acid is obtained from this mixture by the crystallization of the zinc salts from alcohol and purifying the acid by repeated crystallization from ether.

On oxidation it yields a different dihydroxystearic acid than that from oleic acid.

(iii) **Elaidic acid.** $C_{17}H_{33}.CO.OH$.

This is the name given to the substance formed by the action of nitrous acid on oleic acid.

It is a solid which crystallizes from alcohol giving crystals that melt at 44·5° C.

S.G. ·8505 at $\frac{79{\cdot}4^\circ}{4^\circ}$ C.

Its reactions are similar to those of oleic acid and it forms what is known as a **stereo-isomer** of oleic acid; the difference in their constitution may be expressed thus:

Oleic acid

$$\begin{array}{l} CH_3.(CH_2)_7.CH \\ \qquad\qquad\quad \| \\ \qquad\qquad\quad HC.(CH_2)_7.CO.OH \end{array}$$

Elaidic acid

$$\begin{array}{r} CH_3.(CH_2)_7.CH \\ \| \\ HO.OC.(CH_2)_7.CH \end{array}$$

§ **45.** The question of Isomerism has been referred to in § 28. Stereo-Isomerism may be briefly explained as follows:

In the case of oleic acid and elaidic acid it is seen that (1) the number of atoms per molecule is the same in both cases, (2) the arrangement of the atoms in the molecule is the same: and hence the cause of the difference must be sought otherwise. In order to explain the variation in properties of these two compounds it is necessary to consider the arrangement of the valencies of the carbon atom in space. According to the ideas of le Bel and van t' Hoff it is supposed that the valencies or combining forces of the carbon atom are uniformly distributed in space; this will be so if one imagines that the carbon atom is situated at the centre of a regular tetrahedron (*O* Figs. 4 and 5), and that the four valencies are directed towards the four solid angles of that figure, viz. in the figures the angles *AOB*, *BOC*, *COD*, *DOA*, *DOB*, *AOC*, are all equal and hence the valencies *OA*, *OB*, *OC*, *OD* are evenly distributed

[1] Arnand and Posternak, *Compt. Rend.* 1910 (150), 1520.

in space. Now compare Fig. 4 with Fig. 5 and it will be noted that no matter how Fig. 5 is rotated it cannot produce Fig. 4, in fact Fig. 5 is the image of Fig. 4 as seen in a mirror. Imagine one's self standing at *A* in each figure and looking down on the base, then it will be seen that the letters *B*, *C*, *D* in Fig. 4 are in a clockwise direction, whilst in Fig. 5 the same letters are in an anti-clockwise direction. It will thus be seen that though Fig. 4 and Fig. 5 are identical both as regards the structure and letters, the spatial arrangement is different, and therefore Fig. 4 and

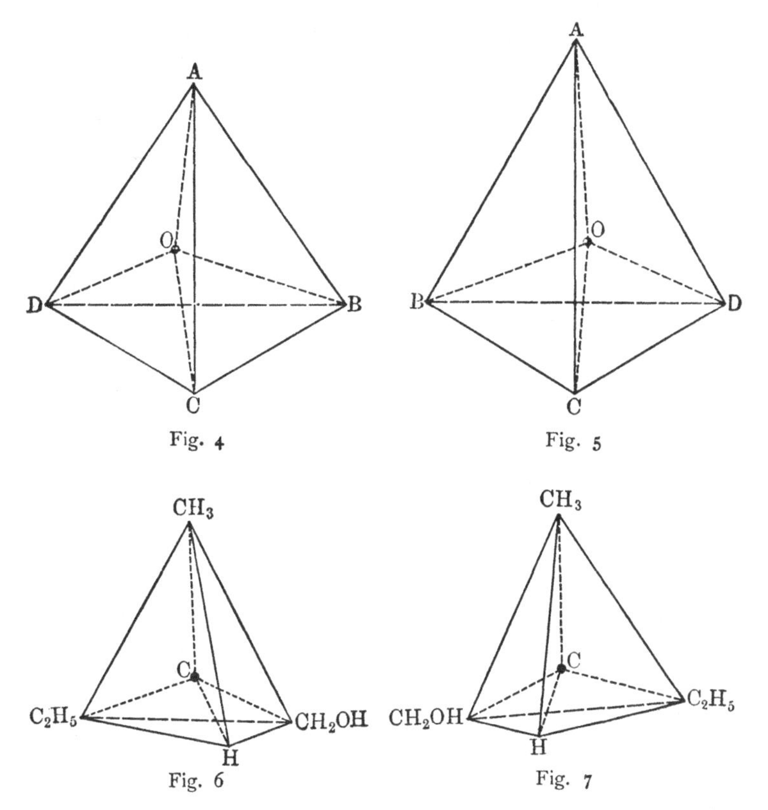

Fig. 4

Fig. 5

Fig. 6

Fig. 7

Fig. 5 represent two different compounds; this only holds good when the letters at the four angles are different, i.e. as long as the atoms or groups situated at the positions occupied by the letters *A*, *B*, *C*, *D* differ each from one another; e.g. the figures 6 and 7 represent the structure of the two different amyl alcohols.

In the compounds mentioned only single valencies occur between the carbon atoms, whilst in oleic and elaidic acid there are double linkages. Now double linkages can be represented diagrammatically as follows:

i.e. two tetrahedra meeting at an edge. In Fig. 8, *A*, *B*, *C*, *D* represent four groups attached to two carbon atoms and also the same in Fig. 9, but it is seen that whereas *A* and *C* are opposite each other in Fig. 8, *A* and *D* are opposite in Fig. 9, hence the grouping in Fig. 8 is spatially

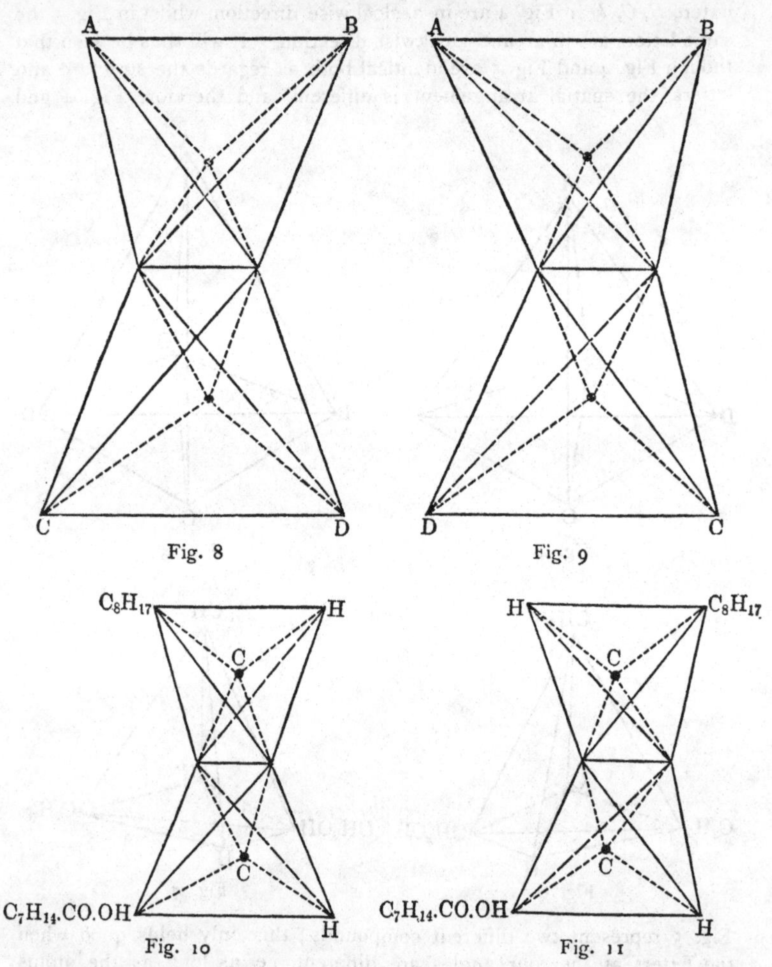

Fig. 8

Fig. 9

Fig. 10

Fig. 11

different from that in Fig. 9, i.e. these arrangements represent two different compounds having the same composition and same structural arrangement. Oleic acid and elaidic acid belong to this type, as shown in figs. 10 and 11.

Thus compounds possessing the same composition and structure, but having a different spatial arrangement, are said to be *stereo-isomeric*.

(iv) **Rapic acid**[1] occurs in rape oil.

(v) **Petroselinic acid**[2] double linking probably 6—7, viz.:

$$\begin{array}{l} CH_3.(CH_2)_{10}.CH \\ \qquad\qquad\qquad\;\; \| \\ \qquad\qquad\qquad CH(CH_2)_4.CO.OH. \quad \text{M.P. } 33^\circ C. \end{array}$$

occurs in Parsley seeds.

(vi) Cheiranthic acid[3] occurs in Cheiranthus oil; M.P. 30° C.

(vii) Liver Lecithin oleic acid[4] from Liver Lecithin.

Acids C_{19}. Two acids of this composition, of doubtful identity, are said to occur in oils.

(i) Doeglic acid[5] in Arctic sperm oil.

(ii) Jecoleic acid[6] in Cod liver oil.

Acids C_{20}.

Gadoleic acid[7] in Cod liver and Whale oils; M.P. 24·5° C.

Acids C_{22}. Of these acids the most important is

Erucic acid. Double linking 13. 14, viz.

$$\begin{array}{l} CH_3(CH_2)_7.\overset{14}{C}H \\ \qquad\qquad\quad\;\; \| \\ \qquad\qquad\quad \underset{13}{C}H.(CH_2)_{11}\ CO.OH. \quad \text{M.P. } 33^\circ\text{—}34^\circ C. \end{array}$$

This acid occurs in rape oil, mustard oils and fish oils, also the oil from tropœlum. Crystallizes from alcohol in needles.

It resembles oleic acid in its chemical reactions very closely; thus nitrous acid converts it into a stereo-isomeride termed:

Brassidic acid. $$\begin{array}{r} CH_3.(CH_2)_7.CH \\ \| \\ COOH.(CH_2)_{11}.CH. \quad \text{M.P. } 65^\circ C. \end{array}$$

analogous to elaidic acid.

On heating with fuming hydriodic acid and phosphorus it is reduced to the corresponding saturated acid, viz. behenic acid.

Melted with caustic potash it splits into acetic and arachidic acids (cf. oleic).

The dibromide on treatment with alcoholic potash yields behenolic acid, a homologue of stearolic acid (q.v.).

§ 46. Acids of the Linolic Series. $C_nH_{2n-3}.CO.OH$.

These acids contain probably TWO PAIRS of DOUBLY LINKED carbon atoms. Hence they absorb FOUR ATOMS of bromine, or four atoms of hydrogen in the presence of a catalyst, being in the latter case reduced to saturated acids.

[1] *Jour. Soc. Chem. Ind.* 1887, 732.
[2] Von Gerichten and Köhler, *Ber.* 1909, 1638.
[3] Matthes and Boltze, *Arch. d. Pharm.* 1912 (250), 211.
[4] Hartley, *Jour. f. Physiol.* 1909 (38), 367.
[5] Scharling, *Jour. f. Prakt. Chem.* 1848 (43), 257.
[6] Heyerdahl, *Cod Liver Oil and Chemistry*, xcviii.
[7] Bull. *Ber.* 1906, 3574.

They readily absorb oxygen on exposure to the air. Nitrous acid does not give 'elaidins.' The lead salts are ether soluble like those of the oleic series.

Two only are definitely proved as occurring naturally, both being C_{18} acids. viz.

Linolic acid $C_{17}H_{31}.CO.OH$. M.P. under $-18°$ C.
Elæomargaric acid " . M.P. 48° C.

Linolic acid. $C_{17}H_{31}.CO.OH$.

This acid occurs in drying and semi-drying oils.

Its constitution has not yet been determined, though Bedford[1] appears to conclude that it is a mixture of two linolic acids, whilst Goldsobel[2] assigns it the formula

$$CH_3.(CH_2)_4.CH:CH.CH_2.CH:CH.(CH_2)_7.CO.OH.$$

It is a water-white oily liquid.

Fluid at $-18°$ C.

S.G. ·9026 at 18° C.

B.P. 229°—230° C. at 16 mm.

Slightly acid reaction.

Readily soluble in alcohol and ether.

Forms a skin in a few days on exposure to air or oxygen.

On bromination it yields a crystalline tetrabromide $C_{18}H_{32}O_2Br_4$ readily soluble in alcohol and ether, but **sparingly soluble in petroleum ether** (cf. dibromide of oleic acid). M.P. 114° C.

A liquid tetrabromide is also produced soluble in petroleum ether [this was recently obtained crystalline from solution in methyl alcohol. M.P. 54°—55°][3].

On oxidation with alkaline permanganate solution a tetrahydroxy stearic acid is produced, viz.

$$C_{17}H_{31}(OH)_4.CO.OH. \quad \text{(Sativic acid.)}$$

Salts of Linolic acid.

Ammonium linolate. M.P. 57°—58° C. sol. in alcohol, CS_2, CCl_4, $CHCl_3$, warm $(CH_3)_2CO$ and C_6H_6.

Potassium " sol. in alcohol, less so in water.

Ca, Ba, Zn, Cu and Pb linolates are soluble in ether.

Barium linolate easily soluble in benzene and petroleum ether (cf. Oleic acid).

Elæomargaric acid (elæostearic acid). $C_{17}H_{31}.CO.OH$.

This is a stereo-isomer of linolic acid (see § 45). It occurs in tung oil.

M.P. 48° C.

Readily absorbs oxygen and dries to a resinous mass.

§ 47. In addition to the Linolic series there are two other series of acids of the same type formula, viz. $C_nH_{2n-3}.CO.OH$, but of totally different constitution; one of these, the **Tariric acid** series contain a **triple linkage** between two

[1] *Inaug. Dissert.*, Halle a/S. 1906.
[2] *Jour. Russ. Phys. Chem. Soc.* 1906 (38), 182.
[3] Matthews and Boltze, *Arch. d. Pharm.* 1912 (250), 225.

carbon atoms, whilst the other, the **Chaulmoogric series** possesses a **cyclic structure.**

Tariric acid. $CH_3 . (CH_2)_{10} . C : C . (CH_2)_4 . CO . OH$[1].

This is found in the oil from the seeds of Tariri, Guatemala.

M.P. 50·5° C.

Absorbs four atoms of bromine.

On reduction yields stearic acid.

Chaulmoogric series. The constitution of the acids of this series is represented by the following **tautomeric** formulae[2].

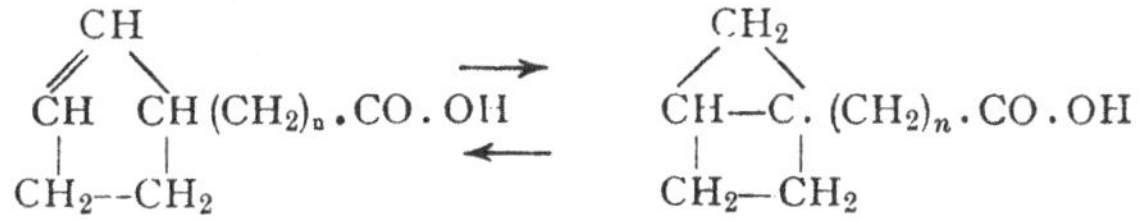

```
     CH                                  CH2
   //  \                   --->         /  \
 CH    CH(CH2)n.CO.OH               CH—C.(CH2)n.CO.OH
  |     |                  <---      |   |
 CH2--CH2                           CH2—CH2
```

They contain only one pair of doubly linked carbon atoms and hence absorb only two atoms of bromine. They are OPTICALLY ACTIVE.

Hyndocarpic acid. $C_{16}H_{28}O_2$.

Occurs in Hyndocarpus and Chaulmoogric oils.

Easily soluble in chloroform; nearly insoluble in other organic solvents.

Chaulmoogric acid. $C_{18}H_{32}O_2$.

Occurs in Chaulmoogra oil.

M.P. 68° C.

§ 48. Acids of the Linolenic Series. $C_nH_{2n-5} . CO . OH$.

The acids of this series probably contain THREE PAIRS of doubly linked carbon atoms. They absorb **six atoms** of Br per molecule and are reduced by H and a catalyst to the corresponding saturated fatty acid. They readily dry when exposed to air by absorption of oxygen.

The lead and barium salts are easily soluble in ether.

Nitrous acid does not convert them into "Elaidins."

Linolenic acid is the only member known to occur in oils.

Linolenic acid. $C_{17}H_{29} . CO . OH$.

Occurs in all vegetable drying oils. Characteristic of linseed oil.

There are probably two linolenic acids, one of which occurs in linseed oil. This one gives a crystalline hexa-bromide. M.P. 180° C. It is a water-white liquid. S.G. ·9046 $\frac{20^\circ}{4^\circ}$ C. It absorbs oxygen rapidly from the air becoming dark brown in colour. Ozone converts it into an *ozonide*.

On oxidation with dilute alkaline permanganate it produces two hexahydroxy stearic acids, viz., linusic and iso-linusic acids

$$C_{17}H_{29}(OH)_6 . CO . OH.$$

Acids of the Clupanodonic Series. $C_nH_{2n-7} . CO . OH$.

Isanic acid[3]. $C_{13}H_{19} . CO . OH$.

Said to occur in seeds of Isano. M.P. 41° C. Existence doubtful.

[1] Arnaud, *Compt. Rend.* 1902 (134), 473, 842.
[2] Barrowcliff and Power, *Jour. Chem. Soc.* 1907, 577.
[3] *Jour. Soc. Chem. Ind.* 1896, 660.

Clupanodonic acid[1]. $C_{17}H_{27}.CO.OH$.

This acid occurs in the fatty acids of Japanese sardine-oil, herring, whale oils and apparently in all oils of marine group.

It is a pale yellow liquid with a fishy odour.

Oxidizes to a varnish-like mass in air.

It absorbs eight atoms of bromine, giving an octo-bromide

$$C_{18}H_{28}O_2Br_8.$$

This is sparingly soluble in alcohol and ether, even on heating; it does not melt below 200° C. (cf. hexabromide of linolenic acid).

With H and a catalyst produces stearic acid.

Arachidonic acid[2]. $C_{19}H_{31}.CO.OH$. The existence of this acid is inferred from the formation of an octobromide of an acid obtained in liver lecithin.

§ 49. Hydroxylated acids.

These acids differ from those already considered inasmuch as they contain one or more hydroxyl (—OH) groups in the molecule. Each hydroxyl group is attached to a different carbon atom, i.e. no carbon atom is combined with more than one hydroxyl group.

The presence of this group in an acid bestows special properties upon the compound, e.g.:

(1) It tends to be more soluble in water than the corresponding non-hydroxy acid whether saturated or unsaturated.

(2) It is also more soluble in alcohol.

(3) A remarkable property possessed by all hydroxy-acids in which the —OH group is attached to the 3rd carbon atom (called the γ carbon atom) from the carboxyl group is their power of eliminating 1 molecule of water and producing an inner anhydride termed a **lactone** viz.

$$\boxed{HO}.\overset{3}{\underset{\gamma}{C}}H_2.\overset{2}{\underset{\beta}{C}}H_2.\overset{1}{\underset{\alpha}{C}}H_2.CO.O\boxed{H} \quad \gamma \text{ or } 3 \text{ hydroxy-butyric acid}$$

eliminates 1 molecule of H_2O, e.g. the —OH group and the H of the carboxyl group producing butyro-lactone viz.

$$\underbrace{CH_2.CH_2.CH_2.CO.O} \text{ or } \begin{matrix} \overset{1}{C}H_2.CO \\ | \\ \underset{2}{C}H_2.\underset{3}{C}H_2 \end{matrix} \rangle O$$

(4) The —OH group in these acids possesses similar properties to those it possesses in the alcohols, namely, it has the power of combining with the acetyl group $CH_3.CO.$ of acetyl chloride e.g.

$$\underset{\text{Ethyl alcohol}}{C_2H_5.O\boxed{H}}+\underset{\text{Acetyl chloride}}{CH_3.CO.\boxed{Cl}}=\underset{\text{Ethyl acetate}}{C_2H_5.O.CO.CH_3}+HCl$$

$$\boxed{H}O.CH_2.CH_2.CH_2.CO.OH+CH_3.CO.\boxed{Cl}$$
$$=CH_3.CO.O.CH_2.CH_2.CH_2.CO.OH+HCl.$$

[1] M. Tsujimoto, *Jour. Coll. of Eng. Tokio Imp. Univ.* 1906, IV. 1, 1908, 5.
[2] Hartley, *Jour. of Physiol.* 1909 (38), 353.

Saturated Hydroxylated acids.

The following three are said to occur naturally.

Sabinic acid[1]. $CH_2(OH).(CH_2)_{10}.CO.OH$.

Occurs in the leaves of Juniperus Sabina.

Sparingly soluble in petroleum ether.

Juniperic acid[2]. $CH_2.(OH).(CH_2)_{14}.CO.OH$.

Occurs as above. M. P. 95° C.

Lanopalmic acid. $C_{16}H_{32}O_3$.

Said to occur in wool-wax.

§ **50. Acids of the Ricinoleic Series.** These are mono-hydroxylated oleic series acids:

Ricinoleic acid. $C_{17}H_{32}.(OH).CO.OH$.

The glyceride of this acid forms a large percentage of castor oil.

It is a thick oil of S.G. ·9509 at 15·5° C.

Solidifies at −6° to −10° C. and melts at 4° to 5° C.

It is miscible with alcohol and ether.

The constitution of the acid is

$$CH_3.(CH_2)_5.\mathbf{C}H(OH).CH_2.CH \\ \| \\ CH.(CH_2)_7.CO.OH.$$

It will be noted that the carbon atom printed in heavy type is linked up to four different groups, viz.

(1) $—CH_3.(CH_2)_5$; (2) H—; (3) HO—; and (4) $—CH_2.CH{=}CH(CH_2)_7.CO.OH$.

Such a carbon atom in a molecule of a compound is said to be **asymmetric.** Let us for the sake of simplicity call these four groups *A*, *B*, *C*, *D*, and as in § 35 let us construct the spatial arrangements of the carbon atom O, viz.

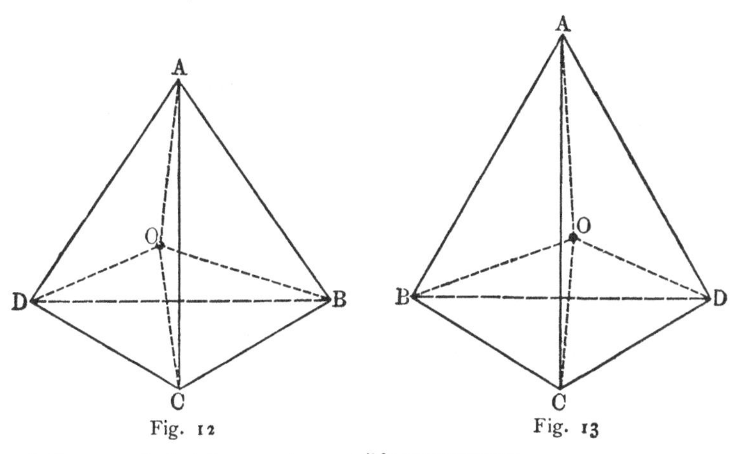

Fig. 12 Fig. 13

as before we note that there are two possible arrangements.

[1] *Compt. Rend.* 1908, 1311; 1910 (150), 875.
[2] Bougault and Bourdier, *Compt. Rend.* 1908, 1311.

Now if we attempt to draw a plane so as to divide either tetrahedron into two equal parts we shall find it impossible to obtain two halves that are similar, no matter how the plane is drawn, but if for one of the groups *B*, *C*, *D* we substitute a group similar to one of the remaining two, viz. substitute *C* by *B* then we obtain only one possible spatial arrangement, viz.

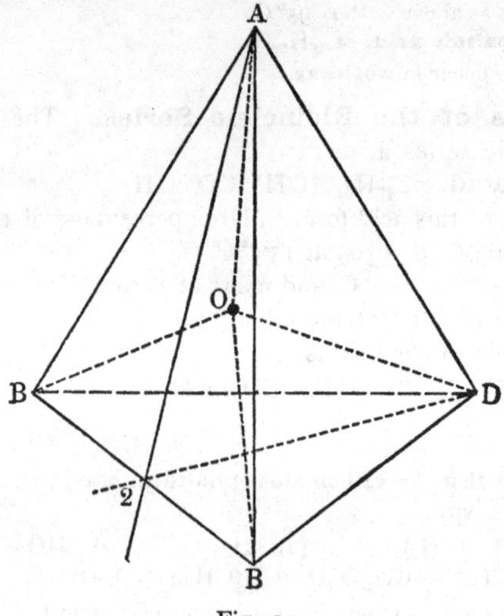

Fig. 14

and now we shall find it possible to draw a plane *A2DA* that cuts the tetrahedron into two similar halves, i.e. the arrangement *ABBD* is symmetrical about *O*, whilst *ABCD* is asymmetrical, hence for asymmetry about a carbon atom it is necessary that the 4 groups attached to it shall be unlike.

Many compounds are known such that if a beam of plane polarized light[1] be sent through a column of it either in the liquid form or in a solution of some solvent, the plane of polarization is rotated through a certain angle, depending upon (1) the length of the column, (2) the strength of the solution, and (3) upon the particular substance; such a compound is said to be "optically active" and such a compound is found to contain an asymmetric carbon atom. Ricinoleic acid is optically active and it contains an asymmetric carbon atom in the molecule. It has been shewn that there are two possible arrangements in space of the four different groups about the asymmetric carbon atom and where the two different compounds are known it is found that whilst one rotates the plane of polarization clockwise the other rotates it anti-clockwise and moreover the rotation is equal in amount for equal lengths of columns of the same strength solutions. One form is thus said to be dextro-rotatory whilst the other form is lævo-rotatory. A solution of equal weights of the dextro- and lævo-compound is optically inactive, i.e. it does not change the direction of the plane of polarization of the light.

Ricinoleic acid is dextro-rotatory $a_D = +6{\cdot}67^\circ$ (100 mm. tube).

[1] Polarized light is light in which the plane of the *direction* of vibration is known.

Ricinoleic acid unites directly with two atoms of Br producing a dibromide.

On reduction at low temperature with H and colloidal palladium, hydroxystearic acid is produced, whilst at higher temperatures and using Ni as the catalyst the —OH group is split off, replaced by H with the formation of stearic acid.

Nitrous acid acts upon it forming the stereo-isomeride **ricinelaidic acid.**

$$\begin{array}{l} CH_3 . (CH_2)_5 . CH . OH . CH_2 . CH \\ \qquad\qquad\qquad\qquad\qquad \| \\ \qquad\qquad CO . OH . (CH_2)_7 . CH. \end{array}$$

M.P. 52°—53° C.

On oxidation with dilute alkaline permanganate two —OH groups are added on producing tri-hydroxystearic acid.

On boiling with acetic anhydride, the **acetyl derivative** is produced, viz.:

$$\begin{array}{l} CH_3 . (CH_2)_5 . CH . (O . \mathbf{CH_3 . CO}) . CH_2 . CH \\ \qquad\qquad\qquad\qquad\qquad\qquad\qquad\quad \| \\ \qquad\qquad\qquad\qquad\qquad\qquad\qquad CH . (CH_2)_7 . CO . OH. \end{array}$$

On treatment of this acetyl compound with NaOH solution in excess, sodium acetate results, and the —OH group is reformed. The amount of acetic acid split off can be estimated: this gives the **acetyl value** (see acetyl value test in practical work).

Concentrated sulphuric acid gives the following products :

Ricinoleosulphuric acid.

$$\begin{array}{l} CH_3 . (CH_2)_5 . CH . (O . HSO_3) . CH_2 . CH \\ \qquad\qquad\qquad\qquad\qquad\qquad\qquad \| \\ \qquad\qquad\qquad\qquad\qquad\qquad CH . (CH_2)_7 . CO . OH. \end{array}$$

Dihydroxystearo-sulphuric acid.

$$\begin{array}{l} CH_3 . (CH_2)_5 . CH(OH) . CH_2 . CH_2 \\ \qquad\qquad\qquad\qquad\qquad\qquad | \\ \qquad\qquad (HSO_3 . O) . CH . (CH_2)_7 . CO . OH. \end{array}$$

Dibasic diricinoleic acid.

$$\begin{array}{l} CH_3 . (CH_2)_5 . CH . CH_2 . CH \\ \qquad\qquad\qquad \diagdown \qquad\quad \| \\ \qquad\qquad\qquad\;\; O \qquad CH . (CH_2)_7 . CO . OH \\ \qquad\qquad\qquad \diagup \\ CH_3 . (CH_2)_5 . CH . CH_2 . CH \\ \qquad\qquad\qquad\qquad\qquad \| \\ \qquad\qquad\qquad\qquad\quad CH . (CH_2)_7 . CO . OH. \end{array}$$

Monobasic, diricinoleic acid.

$$\begin{array}{l} CH_3 . (CH_2)_5 . CH(OH) . CH_2 . CH \\ \qquad\qquad\qquad\qquad\qquad\qquad \| \\ \qquad\qquad\qquad\qquad\qquad CH . CH_2 . (CH_2)_6 . CO \\ \qquad\qquad\qquad\qquad\qquad\qquad\qquad\qquad\qquad \diagdown \\ \qquad\qquad\qquad\qquad\qquad\qquad\qquad\qquad\qquad\;\; O \\ \qquad\qquad\qquad\qquad\qquad\qquad\qquad\qquad\qquad \diagup \\ \qquad\qquad\qquad\qquad\qquad\qquad CH . CH_2 . CH . (CH_2)_5 . CH_3 \\ \qquad\qquad\qquad\qquad\qquad\qquad\; \| \\ \qquad\qquad\qquad HO . OC(CH_2)_7 . CH. \end{array}$$

Dihydroxystearic acid.

$$CH_3 . (CH_2)_5 . CH(OH) . CH_2 . CH_2$$
$$(HO)CH . (CH_2)_7 . COOH.$$

A solid acid. $C_{36}H_{70}O_7$.

Isoricinoleic acid. $C_{17}H_{33} . CO . OH$.

Salts of Ricinoleic acid.

These closely resemble the oleates in their behaviour towards solvents.

Calcium Ricinoleate soluble in warm alcohol.

Barium ,, ,, ,, ,, , insoluble in ether.

Lead ,, easily soluble in ether, insoluble in petroleum ether.

Dihydroxylated acids.

Dihydroxystearic acid. $C_{17}H_{33}(OH)_2 . CO . OH$.

Occurs to a small extent in Castor oil.

M.P. 141° C. Insoluble in ether and petroleum ether; soluble in boiling alcohol.

Lanoceric acid. $C_{29}H_{57}(OH)_2 . CO . OH$.

Occurs in wool wax.

M.P. 104°—105° C. Readily soluble in hot alcohol.

Dibasic acids.

Japanic acid. $C_{19}H_{38} : (CO . OH)_2$.

This acid first identified in oils and fats[1]. Found in Japan wax.

M.P. 117·7°—117·9° C.

§ 51. Oxidation products of fatty acids.

In addition to those hydroxylated acids which occur naturally, other oxidation products, yielding **anhydrides**, are produced in the oils and fats industries, and are therefore of interest. These are formed by the oxidation of unsaturated acids and form a test for the identity of these acids, each pair of doubly-linked carbon atoms absorbing two hydroxyl groups.

[The **Stearolic** group of acids which contain a pair of **trebly-linked** carbon atoms, are of great theoretical interest as representing a stage in the splitting of the double bond, giving products by which the position of the latter in the molecule can be ascertained.]

Hydroxylated acids.

These are obtained by the oxidation of unsaturated fatty acids with alkaline permanganate. **As many OH groups are assimilated as there are unsaturated carbon atoms in the molecule.** (Hazura's Rule.)

The following is a list of some unsaturated acids and their hydroxylated products: (Lewkowitsch).

Unsaturated acid.		**Hydroxylated Derivative.**
Tiglic	acid.	Dihydroxytiglic acid.
Hypogæic	,,	Dihydroxypalmitic ,,
Palmitoleic	,,	Dihydroxypalmitoleic ,,
Chaulmoogric	,,	α, β dihydroxy dihydrochaulmoogric acid.
Oleic	,,	Dihydroxystearic ,,

[1] L. A. Eberhardt, *Inaug. Diss.*, Strasburg, 1888.

Unsaturated acid.		**Hydroxylated Derivative.**		
Elaïdic	acid	Dihydroxystearidic	acid	
"Iso-oleic"	,,	"Paradihydroxystearic"	,,	
Petroselenic	,,	Dihydroxypetroselenic	,,	
Ricinoleic	,,	Trihydroxystearic acid (and an iso- acid)		
Ricinelaidic	,,	Iso-trihydroxystearic acid		
Linolic	,,	Tetrahydroxystearic	,,	(Sativic acid)
Linolenic	,,	Hexahydroxystearic	,,	(Linusic ,,)
Gadoleic	,,	Dihydroxygadoleic	,,	
Erucic	,,	Dihydroxybehenic	,,	
Brassidic	,,	Iso-dihydroxybehenic	,,	
Arachidonic	,,	Octohydroxystearic	,,	
Clupanodonic	,,	Octohydroxystearic	,,	

Monohydroxylated acids.

(1, 10) **Hydroxystearic acid.** $CH_3.(CH_2)_7.CH(OH).(CH_2)_8.CO.OH$.

Prepared commercially by treatment of a petroleum ether solution of oleic acid with concentrated H_2SO_4.

Crystalline body. M.P. 81°—85° C.

Alcohol (at 20° C.) dissolves 8·8 %.

Ether ,, ,, 2·3 %.

On distillation in vacuo, part of the acid changes into oleic and iso-oleic acids.

(1, 11) **Hydroxystearic acid.** $CH_3.(CH_2)_6.CH.(OH).(CH_2)_9.CO.OH$.

Formed in treatment of "iso-oleic" acid with sulphuric acid at low temperatures.

M.P. 77°—79° C.; more soluble in ether than the 1, 10 acid.

But alcohol at 20° C. dissolves only 0·5 %.

Distils unchanged at 100 mm. (cf. 1, 10 acid).

(1, 4) **Hydroxystearic acid.** $CH_3.(CH_2)_{13}.CH(OH).(CH_2)_2.CO.OH$.

The free acid has not been prepared as it passes into an **inner anhydride** or **lactone** by the elimination of one molecule of water, a property common to all 1, 4 (or γ) hydroxy-acids, viz.

$$CH_3.(CH_2)_{13}.\overset{4}{\underset{\gamma}{C}}H\boxed{(OH)}.\overset{3}{\underset{\beta}{C}}H_2.\overset{2}{\underset{\alpha}{C}}H_2.\overset{1}{C}O.O\boxed{H}$$

$$= CH_3(CH_2)_{13}.CH.CH_2.CH_2.CO + H_2O.$$

(the CH and CO of the product are joined through O, forming a ring)

Stearo-lactone as above is one of the products of interaction of concentrated H_2SO_4 on ordinary oleic acid; it is also formed on heating ordinary oleic acid with 10 % of $ZnCl_2$ to 158° C.

M.P. 47°—48° C.; distils unchanged (almost).

Soluble in alcohol, ether and petroleum ether.

Dihydroxylated acids. The only important member is

9, 10 **Dihydroxystearic acid.**

$$CH_3.(CH_2)_7.CH(OH).CH(OH).(CH_2)_7.CO.OH.$$

Prepared by oxidizing ordinary oleic acid with alkaline permanganate.

M.P. 133° C.; soluble in hot alcohol, easily; sparingly in ether; insoluble in water.

[13, 14 Dihydroxybehenic acid. $CH_3(CH_2)_7 . CH . OH$
|
$CH . OH . (CH_2)_{11} . CO . OH$.

Formed on oxidation of Erucic acid.

M.P. 132° C. ; soluble in warm alcohol, insoluble in ether.]

Trihydroxylated acids. These are formed by the oxidation of Ricinoleic acid. There are two forms, both of which are optically active

(1) M.P. 140°—142° C.
(2) M.P. 110°—111° C.

Tetrahydroxylated acids.

Sativic acid. $C_{17}H_{31}(OH)_4 . CO . OH$.

Produced by the oxidation of Linolic acid.

M.P. 173° C.; soluble 1 part in 2000 of boiling water, insoluble in cold water, ether, chloroform, carbon disulphide and benzene;

Soluble in hot alcohol and acetic acid readily.

Hexahydroxylated acids.

Linusic acid. $C_{17}H_{29}(OH)_6 . CO . OH$.

Formed by the oxidation of Linolenic acid.

M.P. 203°—205° C.

More soluble than the preceding acid.

Octohydroxylated acids.

Clupanodonic acid should yield octohydroxystearic acid. This has not been examined.

§ 52. Acids of the Stearolic Series.

These are formed by treating the di-bromo-acids of the oleic acid series with alcoholic potash solution; 2 molecules of hydrobromic acid, HBr, are split off and a triple linking is formed, viz.:

$$CH_3 . (CH_2)_7 . C\boxed{HBr} . C\boxed{HBr} . (CH_2)_7 . CO . OH$$
$$= 2HBr + CH_3 . (CH_2)_7 . C \equiv C . (CH_2)_7 . CO . OH.$$

These acids are **very stable** and do not oxidize on exposure to the air. On treatment with iodine chloride, as in the iodine test, **only one molecule of ICl is absorbed.**

Table of acids of Stearolic series

Name	Constitutional Formula	Obtained from	M.P.
Palmitolic	$CH_3(CH_2)_7 . C \equiv C . (CH_2)_5 . CO . OH$	Hypogæoic Dibromide	42° C.
Stearolic	$CH_3(CH_2)_7 . C \equiv C . (CH_2)_7 . CO . OH$	Oleic dibromide	48° C.
Petroselenolic[1]	$CH_3(CH_2)_{10} . C \equiv C . (CH_2)_4 . CO . OH$	Petroselenic ,,	45° C.
Behenolic	$CH_3(CH_2)_7 . C \equiv C . (CH_2)_{11} . CO . OH$	Erucic ,,	57·5° C.
Ricinostearolic	$CH_3 . (CH_2)_5 . CH(OH) . CH_2 . C \equiv C(CH_2)_7 . CO . OH$	Ricinoleic ,,	51° C.

[1] E. Vongerichten and A. Kohler, *Ber.* 1909 (149), 220.

§ 53. Dibasic acids.

The following dibasic acids are interesting as being products of the splitting of unsaturated acids at the position of double-linkage. They are all soluble in water and can be readily isolated and identified.

Name	Formula	M.P.	Solubility in 100 parts of water at 20° C.
Suberic acid	$C_6H_{12}(CO.OH)_2$	140° C.	0·16 parts
Azelaic ,,	$C_7H_{14}(CO.OH)_2$	106·2° C.	0·24 ,,
Sebacic ,,	$C_8H_{16}(CO.OH)_2$	133° C.	0·10 ,,
Brassylic ,,	$C_{11}H_{22}(CO.OH)_2$	112° C.	0·74 (at 24° C.)

§ 54. Alcohols.

As all natural oils and fats are compounds of glycerine, and contain no other alcohol, the alcohols given in the list below have all (with the exception of glycerine) been isolated from **waxes**. Only those whose identity is established beyond doubt are given in the table.

The term **alcohol** is applied to all compounds derived from hydrocarbons by the replacement of one or more hydrogen atoms with one or more **hydroxyl** groups. Consequently there are many series of alcohols. The alcohols containing only one hydroxyl group per molecule are known as **monohydric alcohols**, whilst these containing two are called **dihydric**; there are also **trihydric**, **tetrahydric**, **pentahydric** and **hexahydric** alcohols. The first member of each series contains as many carbon atoms as hydroxyl groups in the alcohol: thus the first member of

(1) The Monohydric alcohols is $CH_3.OH$ Methyl alcohol
(2) The Dihydric ,, ,, $C_2H_4(OH)_2$ Glycol ,,
(3) The Trihydric ,, ,, $C_3H_5(OH)_3$ Glycerol ,, etc.

since it has been found by experience that **no carbon atom[1] is able to carry more than one hydroxyl group.**

Monohydric Alcohols
or
The Ethyl Alcohol Series. $C_nH_{2n+1}OH$.

Name	Formula	M.P.	Occurrence
Cetyl alcohol	$C_{16}H_{33}.OH$	50° C.	Spermaceti as palmitate
Octodecyl ,,	$C_{18}H_{37}.OH$	59° C.	Spermaceti
Arachyl ,,	$C_{20}H_{41}.OH$	70° C.	Fat of Dermoid cysts
Ceryl ,,	$C_{26}H_{53}.OH$	79° C.	Chinese wax
Melissyl ,,	$C_{30}H_{61}.OH$	88° C. (?)	Beeswax as palmitate

Dihydric Alcohols
or
Glycolic Series. $C_nH_{2n}(OH)_2$.

Name	Formula	M.P.	Occurrence
Anonymous	$C_{24}H_{48}(OH)_2$	103° C.	Carnäuba wax[2]
Cocceryl alcohol	$C_{30}H_{60}(OH)_2$	101°—104° C.	Cochineal wax as coccerate

[1] There are very few exceptions to this.
[2] Stürcke, *Liebig's Ann.* 223, 283.

Trihydric Alcohols
or
Glycerols. $C_nH_{2n-1}(OH)_3$.

Name	Formula	M.P.	Occurrence
Glycerol	$C_3H_5(OH)_3$	crystals melt at 20° C.	All natural oils and fats as glyceroids

Cyclic Alcohols
or
Sterols.

Name	Formula	M.P.	Occurrence
Cholesterol	$C_{27}H_{46}O$	148°—150° C.	Free in all oils and fats from animal sources
Iso-Cholesterol	$C_{27}H_{46}O$	137°—138° C.	Wool fat
Bombicesterol	$C_{22}H_{44}O . H_2O$	148° C.	Chrysalis oil
Phytosterols			All vegetable oils
Sisosterol	$C_{27}H_{46}O$ (?)	135° C. (?)	Maize oil
Brassicasterol	$C_{28}H_{46}O . H_2O$	148° C.	Rape oil
Sigmasterol	$C_{30}H_{48}O . H_2O$ (?)	170° C.	Calabar bean
Coprosterol	$C_{25}H_{44}O$ (?)	98°—100° C.	Sewage fats

Monohydric alcohols.

These occur mostly in **waxes**. They are of a slightly basic nature owing to the presence of the OH group, hence they combine with acids to form salts (ethereal salts or esters), the reaction taking place more readily in the presence of a dehydrating agent viz. H_2SO_4.

The reaction is similar to that already given before with glycerine, e.g.

$$\underset{\text{acetic acid}}{CH_3.CO.O\boxed{H}} + \underset{\text{methyl alcohol}}{CH_3.\boxed{OH}} = \underset{\text{methyl acetate (an ester)}}{CH_3.CO.OCH_3} + \underset{\text{water}}{H_2O}.$$

Heated with **soda-lime** the higher monohydric alcohols give off hydrogen and are converted into salts of fatty acids, e.g.

$$\underset{\text{cetyl alcohol}}{C_{15}H_{31}.CH_2.OH} + NaOH = \underset{\text{sodium palmitate}}{C_{15}H_{31}.CO.ONa} + 2H_2.$$

This reaction is made use of in the estimation of the alcohols.

Cetyl alcohol. $C_{16}H_{33}.OH$.

Occurs in **spermaceti** combined with palmitic acid.

It is a white, tasteless, odourless solid.

M.P. 50° C.; B.P. in vacuo 119° C.

S.G. $\frac{60^\circ}{4^\circ}$ = ·8105.

Insoluble in water, soluble in alcohol, ether and benzene.

Ceryl alcohol. $C_{26}H_{53}OH$.

Occurs as ceryl cerotate in **Chinese wax** and as ceryl palmitate in Opium wax.

Also exists free in wool wax.

Traces are found in beeswax, Japan wax, flax wax and carnäuba wax.

M.P. 79° C.

Heated with soda-lime it yields cerotic acid.

Melissyl alcohol. $C_{30}H_{61}OH$.

Occurs as palmitate in **beeswax**; also in carnäuba wax, sugar cane wax, curcus wax and many oils and fats contain small quantities.

Silky needles. M.P. 85°—88° C.

Nearly insoluble in cold but soluble in hot alcohol, sparingly soluble in petroleum ether and chloroform.

Heated with soda-lime yields melissic acid.

Dihydric alcohols (see table).

Trihydric alcohols.

Glycerol. $C_3H_5(OH)_3$. Structural formula:

$$\begin{array}{l} CH_2 . OH \\ | \\ CH . OH \\ | \\ CH_2 . OH. \end{array}$$

OCCURRENCE.

Occurs as the basic constituent in all oils and fats.

PROPERTIES.

It is a colourless, odourless, viscid liquid with a sweet burning taste.

Neutral to litmus.

On exposure to cold for a long time rhombic crystals are formed. M.P. 20° C.

Absorbs 50 °/₀ of water on exposure to moist air.

Difficult to free from last traces of water (obtained anhydrous by standing over sulphuric acid in vacuo).

S.G. $\frac{15^\circ}{15^\circ}$ = 1·26468 [for S.G. of dilute solutions see Tables at end of book].

B.P. at atmosphere = 290° C. with slight decomposition, below 12 mm. pressure it distils unchanged.

On slowly heating at 160° C. it evaporates without residue; when water solutions are boiled, loss of glycerine occurs after concentration of 50 °/₀ is reached.

SOLUBILITY.

Miscible in all proportions with water and alcohol.

Soluble in a mixture of alcohol and ether; also in acetone.

Sparingly soluble in ether alone (1–500).

Insoluble in chloroform, petroleum ether, carbon disulphide, benzene and in oils and fats.

Glycerol possesses **powerful solvent properties** and many substances dissolve more readily in it than in either water or alcohol.

REACTIONS.

On heating rapidly, it decomposes, producing **acrolein** (odour of burnt fat) and leaving a residue of poly-glycerols.

$$\begin{array}{lcl} CH_2 . \boxed{OH} & & CH_2 \\ | & & \| \\ CH \ . \boxed{OH} & = & CH + 2H_2O \\ | & & | \\ C\boxed{H_2} . OH & & CHO. \end{array}$$

These **polyglycerols** are "concentrated" products of glycerine probably produced by the elimination of water thus:

$$\begin{array}{l} CH_2.OH \\ | \\ CH.OH \\ | \\ CH_2.O\boxed{H} \\ CH_2.\boxed{OH} \\ | \\ CH.OH \\ | \\ CH_2.OH \end{array} = H_2O + \begin{array}{l} CH_2.OH \\ | \\ CH.OH \\ | \\ CH_2 \\ \quad > O \text{ diglycerol.} \\ CH_2 \\ | \\ CH.OH \\ | \\ CH_2.OH \end{array}$$

or generally $nC_3H_8O_3 = mH_2O + C_{3n}H_{8n-2m}O_{3n-m}$.

Glycerol dissolves in **concentrated H_2SO_4** forming

$$C_3H_5(OH)_2O.SO_3H,$$

which is dissociated on boiling with dilute acids.

Caustic alkalies, alkaline earths and **lead oxide** are dissolved forming crystalline compounds termed "**glyceroxides**," which are decomposed by water.

Monosodium glyceroxide. $C_3H_5(OH)_2.ONa$.

Formed by mixing a solution of metallic sodium in absolute alcohol (C_2H_5OH) with glycerine. Crystals are deposited of formula $C_3H_5(OH)_2.ONa + C_2H_5OH$, which are very deliquescent. It is probably an α derivative [1], viz.:

$$CH_2.OH.CH_2.OH.CH_2.ONa.$$

Disodium glyceroxide. $C_3H_5(OH)(ONa)_2$.

Prepared by further treatment of the monosodium compound with sodium in absolute alcohol.

The potassium salts are very similar.

OXIDATION PRODUCTS.

(1) Sulphuric acid solutions of potassium permanganate or potassium dichromate oxidize glycerol completely to CO_2 and H_2O, viz.:

$$2KMnO_4 + 3H_2SO_4 = K_2SO_4 + 2MnSO_4 + 3H_2O + 5O$$
$$C_3H_5(OH)_3 + 7O = 3CO_2 + 4H_2O.$$

(2) Alkaline solutions of permanganate oxidize it to oxalic and carbonic acids, viz.:

$$2KMnO_4 = K_2O + 2MnO_2 + 3O$$
$$C_3H_5(OH)_3 + 6O = H_2C_2O_4 + H_2CO_3 + 2H_2O.$$

(3) Dry potassium permanganate reacts violently with glycerine. These reactions are of analytical importance.

Glycerol combines with acids to form **esters**, the most important being those with the **fatty acids**, forming oils and fats.

[1] Nef.

Sulphuric acid esters.

Three esters have been obtained, viz.:

(1) $CH_2(O.SO_3H).CH(OH).CH_2(OH)$.

(2) $C_3H_5(OH).(O.SO_3H)_2$.

(3) $CH_2(O.SO_3H).CH(O.SO_3H).CH_2(O.SO_3H)$.

Phosphoric acid esters.

A natural ester is **Lecithin**, previously described. Three esters, mono-, di- and tri-, are artificially produced by the interaction of phosphoric acid with glycerine under varying conditions. They have recently been extensively employed as nerve stimulants.

Acetic acid esters.

Mono-, di-, and tri-acetins have been described under glycerides (q.v.). Triacetin is the product formed in the **acetin** method for the estimation of glycerol (q.v.).

Nitric acid esters.

All three nitrates—mono-, di-, and tri-,—have been prepared. The trinitrate, known as **nitro-glycerine**—the well-known high explosive—is the chief commercial compound of glycerine.

Glyceryl trinitrate "Nitroglycerine":

$$\begin{array}{l} CH_2.ONO_2 \\ | \\ CH\ .ONO_2 \\ | \\ CH_2.ONO_2. \end{array}$$

This is prepared by running glycerine into a mixture of nitric and sulphuric acids keeping the whole cool (a highly dangerous experiment).

It is a heavy oily liquid; S.G. 1·60.

Violently explosive under certain conditions.

It is an ingredient of many high explosives.

Dynamite is a mixture of nitroglycerin and kieselguhr.

Blasting gelatine is a solution of nitrocellulose in nitroglycerine.

Arsenious acid.

Glyceryl arsenite.

$$\begin{array}{l} CH_2.O \diagdown \\ | \\ CH\ .O - As \\ | \\ CH_2.O \diagup \end{array}$$

formed by the solution of 1 molecule of As_4O_6 in 2 molecules of glycerine by heating to 250° C.

Used as a battery solid.

M.P. 50° C. Volatile in the vapour of glycerol[1].

§ 55. Sterols.

These compounds contain an alcoholic, i.e., an hydroxyl group, OH. They differ, however, from the aliphatic alcohols in not being oxidized to fatty acids on treatment with soda-lime.

[1] It will be noted that it is not possible to remove As from commercial glycerine by distillation.

Cholesterol and its homologues occur as a characteristic constituent in all animal fats.

The **Phytosterols** are found in all oils and fats of vegetable origin. Hence the recognition and properties of these compounds are of great analytical importance.

The sterols are precipitated by an alcoholic solution of **digitonin**[1] $C_{55}H_{94}O_{28}$; one molecule of the latter combines with one molecule of the sterol, forming an addition product insoluble in water, acetone and ether.

The esters of the sterols do not combine with digitonin.

In the case of oils and fats, separation is not easy on account of the resinous bodies usually accompanying the sterol[2].

Cholesterol.

Occurs, as before stated, in all animal fats and oils, also in wool fat in considerable quantity.

CONSTITUTION: probably a terpene compound of the following formula[3]

$$(CH_3)_2 : CH \,.\, CH_2 \,.\, CH_2 \,.\, \underset{\displaystyle |\atop \displaystyle CH_2 \,.\, CH(OH) \,.\, CH_2.}{C_{17}H_{26}} \,.\, CH{=}CH_2$$

PROPERTIES.

Obtained as crystals (anhydrous) from chloroform and as plates containing 1 molecule of water from alcohol (see chap. XIV for micrograph of crystals). M.P.[4] 148·4° C. to 150·8° C.

Insoluble in water; sparingly soluble in dilute alcohol; easily soluble in ether, carbon disulphide and chloroform.

Optically active $[\alpha]_D^{15^\circ} = 31{\cdot}12^\circ$ (lævo).

Volatile on heating carefully at atmospheric pressure without decomposition.

REACTIONS.

Br gives a dibromide $C_{27}H_{46}OBr_2$.[5]

Heated with soda-lime **no fatty acid** produced.

Soluble in H_2SO_4 producing on heating a hydrocarbon ($C_{26}H_{42}$ (?)) (cf. aliphatic alcohols).

For colour reactions see practical section.

Isocholesterol. $C_{27}H_{46}O$[6] an isomer of cholesterol.

Occurs with the latter in wool-fat.

Crystalline needles. M.P. 137°—138° C.[7]

Optically active $[\alpha]_D = +60^\circ$ (in ethereal solution)[7].

[1] Th. Panzer, *Chem. Zentralbl.*, 1912 (ii), 540.
[2] Lewkowitsch, *Oils and Fats*, 264, vol. I.
[3] Windous, *Ber.* 1908, 2568; 1909, 3770.
[4] Bömer.
[5] Wislicenus and Moldenhauer, *Liebig's Ann.* 146, 175.
[6] Lewkowitsch, *ibid.*
[7] Schulze, *Jour. f. Prakt. Chem.* [2], 7. 163.

Bombicesterol. $C_{26}H_{44}O \cdot H_2O$. Occurs in chrysalis oil. M.P. 148° C.

Phytosterols.

Sitosterol. $C_{27}H_{46}O$ (?). A phytosterol first obtained from wheat and maize-oil. Resembles cholesterol in many respects.

Crystals.

M.P. 138°—143° C. according to purity of preparation.

Optically active $[\alpha]_D^{15} = -23{\cdot}14°$.

Brassicasterol. $C_{28}H_{46}O \cdot H_2O$.

Sigmasterol. $C_{30}H_{48}O \cdot H_2O$.

Coprosterol. $C_{25}H_{40}O$. Formed apparently by the reduction of Cholesterol and Phytosterols.

SECTION III

THE TESTING AND ANALYSIS OF OILS, FATS AND WAXES

§ **56.** A full description of the methods used in the analysis of oils, fats and waxes, and directions for the performance of the same will be given in the second volume of this work. In the present chapter will be indicated the **interpretation** of the results obtained in such tests and the **divergencies** which are exhibited with different oils and fats.

The **object** of the examination is usually either

(1) To identify the particular oil, fat or wax, or

(2) To determine the nature and extent of adulteration, if any, which has taken place.

The **method** used in each case is to compare the results of the tests performed with the known figures obtained with pure specimens of the various oils, fats and waxes.

It must be always borne in mind that **modern adulteration** is in many cases carried out under the direction of men with a special and thorough scientific knowledge of the question, and precautions are therefore taken to avoid detection by the usual tests.

Again, apart altogether from the question of wilful sophistication of an oil or fat, a new problem for the oil chemist has been created by the introduction of **hydrogenated** oils in commerce.

Chemical and physical tests may here give no indication of the natural origin of these substances and in some cases it may be impossible even to distinguish them from natural products.

There are however certain differences in appearance and character which to the practised eye may serve to subject them to suspicion.

The specific colour reactions for cotton seed and apparently for sesamé oil are of no avail on account of the destruction of the colour producing substances present in these oils.

A useful indication here is the recognition of **small traces of the catalyst**, e.g. nickel, used in the hydrogenation, which may be left dissolved in the oil or fat.

The presence of **phytosterol** in a stearine would prove a vegetable origin ; and liver oils might still be recognized by the colour reaction with sulphuric acid.

A **hardened maize oil** might be detected by the presence of lecithin.

Apart from this it would seem to be next to impossible in many cases to discover with certainty the source of an oil hardened by hydrogenation. The reason for this will be obvious.

§ **57.** As the oils, fats and waxes are more or less complex mixtures of chemical compounds, the rational method of analysis would be to **isolate** each constituent and determine its identity and the proportion present.

Unfortunately the methods at present at our disposal are so lengthy and cumbersome and require the use of such large quantities of material, that such isolation as a means of technical analysis is not practicable.

The procedure of isolation of the various constituents of an oil or fat becomes still more complex when it is remembered that these contain for the most part **mixed glycerides**, in which the acidic portion consists of two or three different individuals. Thus, assuming that the isolation has been successfully carried out, the identity of each fatty acid radicle in the glyceride has still to be determined: an exceedingly complicated and lengthy process.

In these circumstances, certain physical and chemical "values" and reactions are obtained, which only in a **few cases** (Reichert-Meissl, percentage glycerol, unsaponifiable matter) aim at a separation of the constituents of the substance.

The consideration of these "values," taken together and compared with those typical of pure specimens of the oils, etc., give, in the case of single substances, fairly conclusive results.

The problem becomes much more difficult in the case of a mixture of two oils, the difficulty increasing when the proportion of one of the constituents is small, or when the oils comprising the mixture are closely allied in properties.

In the case of mixtures of three or more glycerides or waxes, the solution except in special cases[1] becomes a kind of scientific guesswork. It is here that knowledge of the commercial aspect of the question is of invaluable assistance in the interpretation of results.

[1] e.g. a mixture of cocoanut, or palm kernel oil, with castor oil and a drying or semi-drying oil, where definite information is obtained by means of the Reichert-Meissl, acetyl and iodine values respectively; a mixture not likely to occur often in practice.

The chief factors considered here are the question of **relative price**—as a rule a very fluctuating figure—and the **suitability** of the substances for the purpose proposed.

The matter is further discussed at length in a later chapter[1].

§ 58. It is preferable in most cases to employ the original oils for testing without preparatory treatment[2], except to remove water and gross impurities or suspended matter.

In this connection it may be pointed out that care is necessary to obtain a truly representative sample of the bulk of oil. The procedure is given in full in the section on practical work.

Metals are determined by incineration of the oil, etc. and subsequently testing the ash by the usual methods.

[1] See " Interpretation of Results " in practical part of book.
[2] See however "Titer test."

CHAPTER IV

PHYSICAL METHODS OF EXAMINATION

§ **59.** The following is a list of the physical methods of examination which yield information for discriminative purposes.

1. Specific gravity.
2. Melting point.
3. Solidifying point of mixed fatty acids.
4. Refractive index.
5. Viscosity.
6. Solubility.
7. Rotatory power.

Of these the first three yield the most useful figures, and are those generally employed.

The **Refractive power** of an oil is, on account of its ease of determination, coming to the fore and is regarded by some chemists as yielding more useful information for discriminative purposes than any other physical test. Recent work on the *dispersive* power of an oil seems to point to a new and valuable test for discriminative purposes (see practical part of book).

The **Viscosity** is an important figure in the valuation of **lubricating oils.**

The **Solubility** of oils, etc. in various solvents yields valuable information with certain oils; whilst the **rotatory power**, as determined by means of the polarimeter, is normally only of use in the case of castor oil, rosin oils and a few somewhat rare oils, not likely to be found in commercial analyses.

The following table indicates the tests of most value with various oils, etc. for discriminative purposes.

Specific Gravity	Melting point	Solidifying point of mixed fatty acids	Refractive index	Viscosity	Solubility	Rotation
Sperm oil, Tung oil, Castor oil Japan wax Carnäuba wax	All waxes	Most oils and fats, especially Cotton-seed oil, Tallow, Cocoanut oil, Fats generally	Most oils, especially Tung oil	Lubricating oils, e.g. Lard oil, Neat's foot oil, Sperm oil, Castor oil	Rape oil, Castor oil, Butter-fat	Castor oil, Rosin oils, Chaulmoogra group

§ 60. SPECIFIC GRAVITY.

The majority of oils and fats are of much the same density, consequently this figure is not of great value in many cases.

The specific gravity of **unsaturated** glycerides being higher than the corresponding saturated ones, it follows that the drying oils will yield higher S.G.s than the non-drying or semi-drying oils. Exposure to air and consequent **oxidation** increases the S.G. of oils, especially in the case of those containing large amounts of unsaturated acids.

The S.G. is useful in the cases of **liquid and solid waxes**, which are seen to be well outside the usual limit for oils and fats; this also applies to castor and tung oils.

The S.G. of solid fats is best taken at the temperature of boiling water[1], liquid fats at 15·5° C. and waxes by suspension in mixtures of alcohol and water.

Table of Specific Gravities at 15·5° C.

(see opposite)

TECHNICAL ASPECTS.

The determination of the specific gravity affords a rapid and extremely useful method of ascertaining the approximate strength of liquid reagents and aqueous solutions. In works' operations the method in general use is that of the **hydrometer**. Arbitrary scales adapted to the particular range of densities required are widely used. Those of **Twaddel** and **Beaumé** are the best known (see tables at end).

The usual method of preparing solutions of definite strength consists in the addition of water until the liquid tested by means of the hydrometer is of the density equivalent to the percentage required. The strength of glycerine during concentration is roughly found in this manner and checked by the Westphal balance or specific gravity bottle method.

Many liquids are sold on the basis of specific gravity which represents either percentage of water or relative purity, e.g.

Liquid caustic potash 50° Bé (1·536)
Sulphuric acid (B.O.V.) 60° Bé (1·706)

also other mineral acids, as well as most organic solvents.

The specific gravity is, of course, the figure employed for determining the volume which a given weight of liquid will occupy, or vice versa, and its determination therefore is of constant service in works calculations.

Substances are frequently used on account of some physical property which they possess and not for chemical purposes. In this case it is the quantity or **volume** which is of importance. A familiar case in oils and fats work is that of **solvents for fats**. These are generally sold by weight (per ton), hence the sp. gr. is necessary in determining their relative efficiency.

[1] For the sake of uniformity, however, solid fats are taken at 15·5° C.

e.g. Supposing solvent A is offered at £20 per ton and solvent B at £25 and the SP. GR. of $A = 1{\cdot}15$ and that of $B = {\cdot}85$. Then, as for the extraction of a given amount of substance, a certain volume of solvent is necessary, the actual relative prices of these will be

$$A = 20 \times 1{\cdot}15 = £23$$

and
$$B = 25 \times {\cdot}85 = £21.\ 5s.$$

Thus B is actually cheaper than A.

Other examples will suggest themselves, e.g. metals used for tanks, vessels, etc. In this connection aluminium is replacing other metals on account of its low sp. gr. and consequent cheapness.

§ 61. THE MELTING POINT.

Oils and fats do not yield a **clearly defined** melting point. This is for two reasons.

(1) Pure triglycerides have two melting points, the higher being that of the crystalline form (see page 12).

(2) Oils and fats are mixtures of a number of glycerides of differing melting points.

Also, as most oils and fats contain a large proportion of **mixed** glycerides, they do not give the same results as artificial mixtures of triglycerides in the proportion of the fatty acids found.

The attention of the student may here be called to the well-known fact that the melting point of a mixture of pure substances is not a mean or average of the components. Consequently, no calculation can be based on this assumption. This is clearly shown in the following table of melting points of mixtures of the pure glycerides tripalmitin and tristearin[1].

Tripalmitin	Tristearin	M.P. °C.
0·0	100·0	56·0[2]
10·0	90·0	60·4
12·0	88·0	60·1
25·0	75·0	58·0
30·6	69·4	57·8
39·8	60·2	56·0
47·0	53·0	57·2
50·0	50·0	56·2
56·2	43·8	55·1
68·8	31·2	54·5
91·6	8·4	60·4
100·0	0·0	62·6

Ubbelohde's modification of Pohl's method is best for commercial work, in which the temperature at which a drop of melted fat forms is taken as the end temperature.

[1] R. Kremann and R. Schoulz, *Monat. f. Chem.*, 1912 (33), 1063.
[2] Lower M.P. See page 12.

The **waxes** yield more definite figures which are important as guides to purity. The ordinary capillary tube method is suitable in this case.

Owing to the **indefinite** nature of the results obtained, the melting point is not of much value in discriminating between various oils and fats. It is however frequently useful in the case of commercial samples of oil containing much or little "stearine" and harder fats such as Japan wax, cacao butter, etc.

§ 62. SOLIDIFYING POINT OF MIXED FATTY ACIDS. "Titer test."

The solidifying point of the mixed fatty acids obtained on saponification and acidifying of oils and fats is a valuable figure for analytical purposes. The method of **Dalican** now usually adopted yields very constant results.

Liquids during solidifying evolve heat (the latent heat of fusion) and a rise of temperature occurs. This rise, not marked with fats, is usually quite distinct with the fatty acids and the temperature taken is the top point of the rise. The table opposite shows the results with some fats:

TECHNICAL ASPECTS.

The temperature at which an oil or fat becomes a solid or liquid respectively is of great importance commercially. In most cases the melting and solidifying points of oils may be lowered by separating a portion of the glycerides, whilst fats may be hardened by crystallisation, and expression of the liquid constituents. Thus 'stearines' are natural fats from which fluid portions have been removed (the separated oil being termed an "Oleine"). The relative hardness[1] of oils, fats and waxes, which is approximately equivalent to the melting point, has special application in the cases following.

1. **Oils.**

(i) **Edible or Salad oils.** These oils should have a sufficiently low S.P. to remain fluid and clear at the indoor winter temperature experienced in the particular country where consumed. The same remark applies to oils used in medical and pharmaceutical practice.

(ii) **Lubricating oils.** If oils of this class solidify during use, there is grave danger of the bearing or moving part to be lubricated running dry owing to the stoppage of flow; consequently this is an important factor and the degree of temperature to which the oil is likely to be exposed must be taken into account.

[1] As by far the bulk of the civilized peoples inhabit countries having temperate climates, these considerations are based upon the normal temperatures obtaining in such latitudes. It is obvious that in tropical and sub-tropical climates many "fats" are normally "oils" whilst hard fats are of a buttery consistency. Hence the names "palm oil" and "cocoanut oil" to the bodies which are liquids in the country of origin.

(iii) **Illuminating oils.** The same remark applies here as in the case of lubricating oils, as the ascent of the oil in the wick is dependent upon its liquidity at low temperatures.

(iv) **Soap manufacture.**

(*a*) **Hard soap.** Here the hardness of the soap produced is roughly proportional to the "titer" or solidifying point of the mixed fatty acids. Generally speaking the higher the titer of the oil the more valuable it becomes for this purpose, as its use requires the addition of less of the more expensive fats to produce a soap of given consistency.

(*b*) **Soft soap.** The point here is the liability of soaps made from oils of high titer to become cloudy and opaque when exposed to low temperatures. This is due to the crystallization of the potassium salts of the solid fatty acids, termed in commerce "figging." The crystals usually occur as spherical or star-shaped opaque granules of varying size in the clear soap. Much difference of opinion exists as to the desirability of "figging" in soft soaps, but a soap which is quite opaque through excess of "figging" is usually regarded unfavourably by the general public.

(*c*) **Candle making.** In this industry the value of an oil would depend on the titer of the fatty acids yielded by decomposition of the oil, or in other words on the percentage of "stearine" or "candle material" which it was capable of producing.

2. **Fats.**

(i) **Edible fats.** The object here is to obtain as nearly as possible **uniformity** of consistency throughout the year. This is best accomplished by mixing or "blending" of varying proportions of (*a*) individual fats of different origins and degrees of hardness (as in butter), (*b*) different fats and oils of varying hardness and M.P.

(ii) **Lubricating fats and greases.** The same remarks apply as in the case of edible fats, uniformity of consistence throughout the year or in varying conditions of climate, being the desideratum.

(iii) **Soap manufacture.**

(*a*) **Hard soap.** According to the degree of hardness required so varying proportions of fats of high titer are used. Thus **curd soaps** contain large proportions of fats giving fatty acids of higher S.P. than the "washer" soaps in which large proportions of oils are used.

It is a well-known fact that a certain amount of oil[1] in the composition of a hard soap is necessary to produce a free lather. (Cocoanut oil can also be used for this purpose.)

(iv) **Candle manufacture.** From the point of view of this industry the higher the titer and consequently the greater the yield of candle material or "stearine" (hard solid fatty acids) the more valuable is the fat. The suitability of a fat for use as candle material in varying climates is determined by the titer test.

[1] Rosin is also used for this purpose as well as to cheapen the soap.

3. **Waxes.**

The melting point is the most important figure in regard to the technical employment of waxes. Thus in the manufacture of Polishes (boot, furniture, floor, etc.) the brilliancy of the polish yielded by any wax is generally proportional to its hardness and consequently to its melting point.

N.B. It should be noted that many apparently genuine waxes of high melting point contain proportions of **rosin** to artificially raise the figure. Rosin has but a low polishing value on account of its "stickiness" and is consequently generally undesirable in polishes.

It does not however, by any means, follow that two waxes having identical melting points are equal in polishing value, e.g. "myrtle wax" (a glyceride) M.P. 40° C. has a vastly higher polishing value than a paraffin wax of the same M.P.

Mineral waxes are usually graded in value in proportion to their M.PS.

§ 63. REFRACTIVE POWER.

This figure, determined on an instrument known as a refractometer, is useful both as a rapid sorting test and combined with other data, as a discriminating factor in analysis. Of the two usual types of instruments employed, the more usual in fat analysis is the **Zeiss Butyro-refractometer** which gives arbitrary scale readings, convertible into refractive indices on reference to a table supplied.

This is as follows:

Scale divisions	n_D	differences.
0	1·4220	
10	1·4300	8·0
20	1·4377	7·7
30	1·4452	7·5
40	1·4524	7·2
50	1·4593	6·9
60	1·4659	6·6
70	1·4723	6·4
80	1·4783	6·0
90	1·4840	5·7
100	1·4895	5·5

The **Abbé** instrument reads directly the refractive index of the oil or fat examined.

Many **different temperatures** have been used in determinations, but it is very desirable to keep to **one definite figure**; **40° C.**[1] would appear to be a very suitable temperature for the purpose.

[1] In the case of those fats which are solid at 40° C., the determination can be made at a higher temperature and value at 40° C. calculated from it.

There are several other types of refractometers, including the differential instrument of *Jean* and *Amagat*.

The Zeiss and Abbé instruments however meet all demands in this class of work and are very convenient in use, requiring only a drop of the liquid for examination.

Utz has shown that after heating the refractive power of oils is increased, especially in the case of the drying and the semi-drying oils.

The appended table gives the figures for the principal oils, fats and waxes.

TECHNICAL ASPECTS.

This figure, apart from its discriminative value, has, at present, little direct bearing upon technical work.

§ **64. VISCOSITY.** The rate at which liquids flow when acted upon by a force, such as that of gravitation, is dependent upon a property called **viscosity**. It may be defined as the resistance which the **particles** of a liquid offer to their passing the one over the other and is therefore a measure of the **internal friction** of the fluid.

Several forms of measuring instruments have been used for determining the relative viscosity of liquids. These may be summarized as follows:

(1) An apparatus based upon the rate of ascent of air bubbles in the liquid.

(2) An instrument[1] in which a disc is suspended by a wire in a liquid. The wire is put under torsion and the amplitude of the swing or vibration of the disc noted. This is proportionally retarded with the viscosity of the liquid.

(3) An arrangement whereby the speed of rotation of paddles, immersed in the liquid and actuated by means of a falling weight, is measured.

(4) An apparatus for timing the rate of flow of a certain volume of the liquid through a standard orifice.

These instruments, which are termed **viscometers** (or viscosimeters) are for commercial testing and the figures obtained are quite arbitrary, and useful only for comparative tests. Most of the instruments in use belong to class (4), such as Redwood's, Engler's, etc. The **absolute viscosity** of a liquid can be calculated from Poiseuille's formula by timing the rate of flow through capillary tubes. The instrument used[2] is termed the **absolute viscometer** and the designers have obtained figures for various densities of glycerine. It is therefore possible by reference to these results (see tables at end of book) to obtain the viscosity in absolute terms from any efflux instrument. For this purpose the rate

[1] Doolittle's Torsion Viscometer, *J. Amer. Chem. Soc.* XV (1893), 173.

[2] Archbutt and Deeley.

of flow of glycerine solutions of varying densities is found in the particular viscometer chosen and a curve is constructed showing the relation between the time of flow and the absolute viscosity of the solutions used. The instrument thus standardized can be used for obtaining results in absolute measure[1].

Poiseuille's formula is:

$$\eta = \frac{\pi g \partial h r^4 t}{8 V a},$$

where

g = acceleration due to gravity,
∂ = density of the liquid,
h = mean head of the liquid,
t = time of flow,
V = volume of liquid discharged,
r = radius of tube,
a = height of column of liquid.

η = absolute viscosity measured in dynes per sq. cm. if C.G.S. units are used. η may be defined as the force, in dynes, required to move two surfaces of liquid, each 1 sq. cm. in area and distant from each other 1 cm., in opposite directions with a velocity of 1 cm. per sec.

In order to obtain whole numbers the authors propose that the absolute viscosity figure be multiplied by 100 thus ($\eta \times 100$).

The viscosity is of little value in the discrimination of natural oils and fats as there is not sufficient variation between them. The exceptions are the cases of castor, rape, and sperm oils, the first two of which are higher and the last lower than the normal. The figure however is of great value in the **examination of lubricants.**

Absolute Viscosities[2] (Archbutt and Deeley).

($\eta \times 100$)

Water at 68° F. = 1·028.

	60° F. (15·5° C.)	100° F. (37·7° C.)	150° F. (65·5° C.)	212° F. (100° C.)
Sperm oil	42·0	18·5	8·5	4·6
Olive oil	100·8	37·7	15·4	7·0
Rape oil (1)	111·8	42·2	17·7	8·0
" " (2)	117·6	44·8	18·8	8·5
Castor oil	—	272·9	60·5	16·9

TECHNICAL ASPECTS.

As previously stated, this determination is chiefly of use in deciding the relative suitability of lubricating oils for various purposes.

[1] See also Savil and Cox, *Jour. Soc. Chem. Ind.* 1916, **35**, 151.

[2] Although the viscosity of an oil at varying degrees of temperature yields important practical information, it would be desirable to choose a suitable definite temperature for ordinary comparisons; e.g. 40° C. In order to record the fall in viscosity of oils on heating the authors propose a "Viscosity Ratio number," being $\frac{\eta \text{ of oil at } 40^\circ \text{ C.}}{\eta \text{ of oil at } 100^\circ \text{ C.}}$. (See practical section, Vol. II.)

It is sometimes necessary to calculate the time which liquids will take to flow under a given pressure through certain lengths of pipe. This figure is roughly dependent on the viscosity of the liquid and can be calculated therefrom.

§ 65. SOLUBILITY.

The natural oils and fats are with the exception of the **castor oil group** soluble in every proportion in the following solvents :

Ether $(C_2H_5)_2O$	Petroleum Ether C_nH_{2n+2}
Carbon Disulphide CS_2	Chloroform $CHCl_3$
Carbon Tetrachloride CCl_4	Benzene C_6H_6

and some other organic solvents.

They are only partially soluble at ordinary temperatures (with the exception again of castor oil) in :

Acetic Acid $CH_3.CO.OH$	Acetone $(CH_3)_2CO$
Alcohol $C_2H_5.OH$	Phenol C_6H_5OH

The solubilities of each oil and fat vary to some extent.

Valenta and others have determined the temperature at which solution occurs in these solvents by heating equal quantities of each, until a solution is obtained, and then cooling until the point at which **turbidity** occurs.

Jean has also determined the amount dissolved by acetic acid at 50° C.

The authors have shown that **very variable results** are obtained with commercial samples of oils unless free fatty acids (which are soluble in the cold in these solvents) are absent, and glacial acetic acid of precisely the same density is always used. In order to overcome these difficulties they propose[1] to determine the acid value of the oil concurrently with this test and to refer to a table showing the degree of influence of the fatty acid present on the figure obtained. They also propose to standardize the acetic acid employed against a selected oil of known purity. (See practical section of book.)

TECHNICAL ASPECTS.

The question of the degree of solubility of castor oil and blown oils in mineral oils is a very important one in the case of compound lubricants.

§ 66. ROTATORY POWER.

This property is due, as previously explained, to the existence of an asymmetric carbon atom in the molecule.

[1] The "Valenta" determinations (0° C. turbidity figures, using acetic acid) given under the descriptions of the various oils and fats, are those obtained by the authors, and are standardized by reference to neutral almond oil (=80·0° C.). They are corrected for free acidity and we have termed the figures "true Valenta."

For the ratio $\frac{V\times 10}{80}$, and for details of the new Ethyl-Amyl alcohol reagent (Almond Oil=70·0° C.) the reader is referred to the Practical part (Vol. II). The figure for waxes, using the alcohol reagent, would appear to have special value.

It appears to have some bearing on the physiological action of oils, those oils containing notable proportion of these glycerides having either pronounced medicinal properties, or being actually poisonous in character.

Consequently a determination of this figure is of value in deciding the suitability of oils for edible purposes, and in toxicological analysis.

§ 67. FLASH POINT.

This is defined as the temperature at which a substance gives off a vapour forming an inflammable mixture with air. The test is applied by heating the substance in an open vessel (**open test**), or in a closed chamber (**close test**); no special apparatus is necessary in the first case. The "close test" is made by periodically opening the chamber enclosing the oil and applying a light; this is conveniently accomplished by a mechanical contrivance in the **Abel** and **Pensky-Martens** instruments. The Abel apparatus, heated with a water-jacket, is used for determinations of oils with flash points not exceeding 100° C., while the Pensky-Martens is used for oils and fats and mineral lubricating oils which flash at much higher temperatures.

TECHNICAL ASPECTS.

The test is not of much value for general discriminative purposes. It is chiefly of value in deciding whether illuminating oils are free from danger of explosion; the legal limit for the flash point of these oils is 73° F.

It is also of considerable importance in the testing of mineral lubricating oils.

CHAPTER V

CHEMICAL METHODS OF EXAMINATION

§ **68. Unsaturation.** As oils and fats differ chiefly in the amount of **unsaturated fatty acids** which they yield, the most discriminative results are obtained by a determination of this figure. This is found in practice by the percentage of **iodine chloride** in terms of **iodine** which is capable of being absorbed by the substance. The test is termed the **iodine value**.

Depending also, indirectly, on the amount of unsaturated bodies present, but not yielding results sufficiently accurate for quantitative purposes are the following tests:

(1) The bromine thermal test

(2) The sulphuric acid thermal test (Maumené reaction)

(3) The oxygen absorption test (Livache)

(4) The sulphur chloride test.

Only in the case of the oxygen absorption test for drying oils can these be considered as giving any information which is not better afforded by the iodine value. They are however retained here as being somewhat easier of performance by persons without special training and as having still a certain vogue in technical work.

Other tests closely related to the iodine value are:

(1) The insoluble bromide value (" Hexabromide test")

(2) The elaidin test.

They yield information as to the **character** of the unsaturated glycerides present: the insoluble bromide value as to the amount of these glycerides having 6 and 8 unsaturated carbon atoms (e.g. Linolenin, Clupanodonin) whilst the elaidin test indicates the presence or absence of olein glycerides yielding acids with 2 unsaturated carbon atoms in the molecule.

Molecular weight. The molecular weight of the glycerides (which generally implies the **number of carbon atoms** present in the fatty acids derived therefrom) is indicated in the **saponification value** [1]. In this test the glycerol is displaced by an alkali base, and the amount entering into combination with the fatty acid estimated. As only certain classes of oils contain any considerable proportion of glycerides

[1] Assuming the elimination of unsaponifiable matter.

yielding acids having more or less carbon atoms than 16 to 20 (but preponderantly 18) this test has only a limited discriminative value[1].

It is useful technically as shewing:

(1) The **amount of soap** obtainable on saponification;

(2) The percentage of **glycerol** yielded.

For this calculation it is also necessary to know the free acidity (if any), the amount of unsaponifiable matter and if mono- or di-glycerides are present.

The percentage of glycerides of fatty acids of **low molecular weight** which are **volatile in steam** and **soluble in water** (A), or **volatile in steam and insoluble in water** (B), is indicated respectively by the

(A) **Reichert-Meissl value** and

(B) **Polenske value**

and their modifications.

Glycerides of this character are contained in two or three important groups of oils and fats.

Hydroxylated acids. Oils containing glycerides of **hydroxylated fatty acids** are recognized by the **acetyl value**, i.e., by the determination of the number of acetyl groups ($CH_3 . CO$) that can replace the H of the -OH groups. Of these **castor oil** is the only important member. The acetyl value is also of use in determining the percentage of **free alcohols** in waxes.

§ **69.** SPECIAL TESTS. Certain tests and reactions have reference to individual oils and fats, as under:

(1) **Renard's** tests for arachis oil, which is quantitative.

(2) **Tortelli** and **Fortini's** test for rape oil.

(3) **Bellier's** test for arachis oil.

The following are special colour tests, viz.:

(4) Cotton-seed oil—**Halphen's, Becchi's**, and the **nitric acid** test.

(5) Sesamé oil—**Baudouin's test.**

(6) Liver oils—sulphuric acid test.

In addition to these there are some tests for the less important oils which are given under the description of the individual oil or fat.

§ **70.** The foregoing tests are all of use for the actual discrimination of an oil or fat. The following reactions are indicative mainly of the **condition** of the substance, e.g., age, method of refinement, etc. They are of value for identification purposes when considered in relation to the usual condition of oils in commerce. They include:

(1) The **acid value**, or the amount of free fatty acid present (when the character of the acid is known).

(2) The **glycerol content**.

[1] The value is important in the analysis of waxes.

(3) The **unsaponifiable matter.**
(4) The **mono-** and **di-glycerides.**
(5) The percentage of **oxidized acids.**
(6) The percentage of **anhydrides** in the mixed fatty acids from oils.

§ 71. Separation, recognition and determination of the components of oils, fats and waxes.

This includes the following operations:

(1) Lead-salt of the fatty acids soluble in ether. Used for separation of liquid and solid (saturated and unsaturated) fatty acids.

(2) Separation and estimation of the saturated acids, viz.: palmitic, stearic, etc.

(3) Separation and isolation of unsaturated acids, viz.: oleic, linolic, linolenic, clupanodonic, etc.

(4) Isolation and estimation of the sterols. This is of great value in the determination of the source (animal or plant) of an oil or fat.

(5) Determination of the free alcohols in a wax.

§ 72. THE IODINE VALUE.

It has already been shown that unsaturated fatty acids and their glycerides containing doubly linked carbon atoms, absorb 2 atoms of bromine for each double bond in the molecule, viz.:

$$>C=C< \quad \text{becomes} \quad >\underset{\text{Br}}{\underset{|}{C}}-\underset{\text{Br}}{\underset{|}{C}}<$$

Iodine in solution, however, is not absorbed regularly by the same bodies.

The Wijs-Hübl reaction consists in the treatment of the fat with a solution of iodine monochloride ICl. Under these conditions one atom of each halogen is absorbed giving a saturated chlor-iodo product, viz.:

$$>C=C< \quad \text{becomes} \quad >\underset{\text{Cl}}{\underset{|}{C}}-\underset{\text{I}}{\underset{|}{C}}<$$

It should be here noticed that acids and glycerides of the **Stearolic** series, which contain trebly linked carbon atoms, absorb only two halogen atoms in the iodine test and therefore behave like doubly linked carbon atoms, viz.:

$$-C\equiv C- \quad \text{becomes} \quad -\underset{\text{Cl}}{\underset{|}{C}}=\underset{\text{I}}{\underset{|}{C}}-$$

This being so, it is possible to calculate the iodine absorption for each unsaturated fatty acid and pure glyceride and the practical results obtained agree very closely with these figures. Perfectly pure saturated fatty acids and glycerides have **no iodine value.**

A reference table of the theoretical iodine values of unsaturated fatty acids and of mono-, di-, and triglycerides is given in the tables at end of book.

At first sight it may seem strange that compounds containing doubly and trebly linked carbon atoms react with Br, I, etc., whilst singly linked carbon atoms do not. It must not be supposed that a double bond implies that two carbon atoms are combined together more firmly than two carbon atoms united by a single bond. Baeyer has put forward a theory, known as 'Baeyer's Strain Theory,' which explains why doubly and trebly linked carbon atoms are more reactive than singly linked carbon atoms. It has already been shown that the valencies of the carbon atom are distributed in space equally (see § 45). Now when two carbon atoms combine together the force or forces holding the two carbon atoms tend to set themselves in straight lines; this is only possible when the union takes place between **one force** of each atom, viz.:

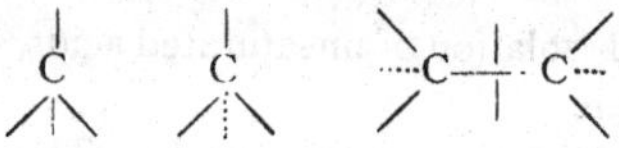

C C —C—C—

whilst if two forces of each atom act upon each other, these forces **tend** to form **straight lines** but in doing so are bent out of their original directions, i.e. they are subjected to a **strain**, viz.:

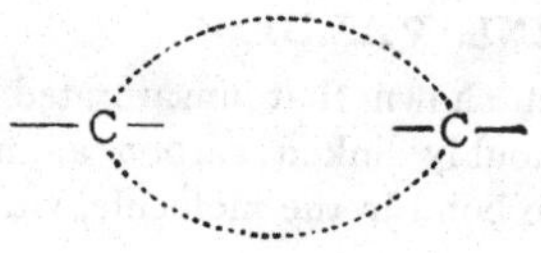

—C— —C—

and in this condition are easily broken apart either entirely or so that one force of each atom is liberated whilst the other two arrange themselves in a straight line, viz.:

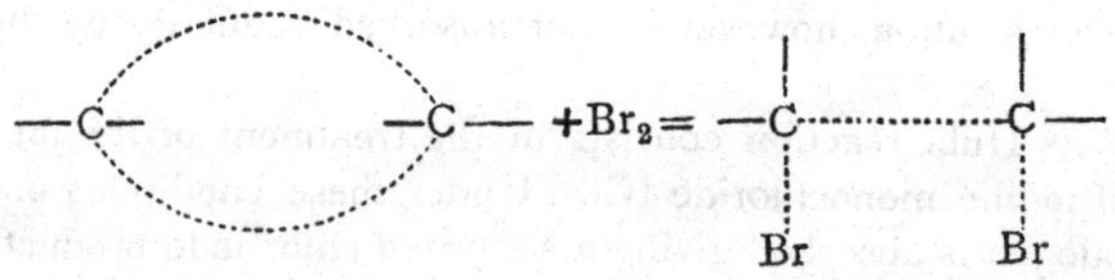

$$-C- \quad -C- + Br_2 = -\overset{|}{\underset{Br}{C}}\cdots\cdots\overset{|}{\underset{Br}{C}}-$$

A table of Iodine Values of more important oils, fats and waxes is given opposite.

§ **73.** TECHNICAL ASPECTS.

The question of the amount of unsaturated glycerides present in an oil has a very important bearing in the case of many industrial applications. These may be tabulated as follows:

1. **Paints and varnishes.**

The **drying power** of an oil is approximately indicated by the iodine value. Oils are thus divided into three main groups.

A. Drying oils.
B. Semi-drying oils.
C. Non-drying oils.

It should however be noted that the iodine value gives no indication of the **character** of the unsaturated glycerides present. For example, **Tung oil** contains little or no linolenic (or isomeric) glyceride, and yet has an iodine value approaching that of linseed oil.

Complete information as to the drying properties of an oil is however not yielded by this test. Account must be taken of the **character of the film** or surface obtained, its elasticity, hardness, degree of 'tackiness' and opacity.

The glyceride of clupanodonic acid, although more highly unsaturated than that of linolenic acid, is not nearly as valuable as a drying oil: hence the inferiority of fish oils over linseed oils and the vegetable oils, for use in paints and varnishes. It should also be noted that oils which have been 'boiled' although reduced in iodine value, are usually better dryers.

2. **Lubricating oils.** Oils containing percentages of the more unsaturated glycerides are liable to thickening and so-called "gumming." The iodine value is therefore of great service in selecting those oils which are free from "gumming" tendencies. Generally speaking oils of low iodine value are suitable as lubricants. The exception is in the case of "blown oils" in which the iodine value has been reduced and the "gumming" tendencies are more or less eliminated.

3. **Burning oils.**

The remarks on lubricating oils apply here also, as oils which "gum" and become thick tend to clog up the wick and are thus quite unsuitable.

In the case of the drying oils, there would be actual danger of spontaneous firing, so violent is the chemical action when the oil is absorbed in the porous fibres of the wick.

4. **Edible oils.**

Oils with high iodine values would here be unsuitable on account of "gumming" properties. They are also usually stronger flavoured than the non-dryers. Many drying oils are however used as edible oils in their country of origin and by the native population. [The figure is of great value in ascertaining the freedom of olive oils, lard, etc. from cheaper semi-drying oils.]

5. **Soap-making.**

Oils of high iodine value have long been used in soft soap making for which they seem quite satisfactory. Generally speaking, oils of whatever iodine value are equally suitable for soap making purposes.

It should, however, be noted that the pure unsaturated glycerides, when saponified, cannot be weakened with water to the same extent as the oils of low iodine value without loss of necessary firmness.

6. **Candle-making.**

The iodine value is an indication of the purity of the stearines used in the industry, and the efficiency of the methods used for the expression of the lower melting point constituents. The products of the *acid process* including "iso-oleic acid" which gives a high iodine value can also be distinguished thus. In the case of artificial stearines produced by hydrogenation the iodine value shows to what extent the hardening has been carried out.

7. **Hardening of oils.** The iodine value beomes an important test in the technical hardening of fats, giving a direct indication of the degree of saturation which has taken place.

It is of course evident that the properties of any oil of high iodine value can be so modified by a greater or less degree of hydrogenation as to become suitable for purposes for which it was originally not fitted.

§ 74. Reactions depending upon the presence of unsaturated glycerides.

1. **The Bromine Thermal Test.**

During the absorption of bromine by an unsaturated fat or fatty acid heat is liberated, the amount being approximately proportional to the percentage of halogen absorbed.

The test is based upon this fact and forms a fairly close guide to the iodine value itself. It is very useful as a rapid sorting test, being capable of performance in a few minutes. The results obtained are shewn in the adjoining table:

2. **The sulphuric acid thermal test** (*Maumené test*).

As a guide to the amount of unsaturated acids present this test is inferior to the preceding (viz. the Br thermal test). The results do not shew such a constant relationship with the iodine value. The test is also easily vitiated by very slight alterations in the strength of the acid[1].

Figures are given in the practical section of the book.

3. **The sulphur chloride test.**

This test, proposed by Fawsett[2], is based upon the fact that sulphur chloride S_2Cl_2 is absorbed by oils or fats in proportion to the amount of unsaturated fatty acids present with the evolution of heat. As a discriminative test it is even less satisfactory than the Maumené test. This is largely because of the time taken for the completion of the reaction, with consequent loss of heat by radiation. The thermal test is therefore not described here. Of more value is the character of the product formed, that with

liquid waxes,
vegetable fats,
animal ,, } being soluble in CS_2,

whilst the product formed with **oils** is only partially soluble in that solvent. This would suggest a ready means[3] of distinguishing between, for instance, sperm oil and neatsfoot oil.

[1] Mazzaron (*J. Agric. Intell. and Plant Diseases*, 1915, **6**, 1700) proposes as a new test the estimation of the SO_2 evolved in this reaction.

[2] *Jour. Soc. Chem. Ind.*, 1882, 552.

[3] This test has not been considered to be of sufficient value to be included with the other "analytical data," p. 94.

4. **The oxygen absorption test** (*Livache*).

This test is of importance practically as shewing the value of an oil for use, as in paints and varnishes. Although in the main giving results comparable with the iodine values, there are some important differences; e.g. **fish and liver oils**, although absorbing as much iodine as the **vegetable drying oils**, have very much less oxygen absorbing power. The test[1] should always be combined with an observation of the **character** of the film formed by drying a thin layer of the oil on a glass plate.

Fish oils do not yield a hard flexible skin as does linseed oil.

5. **The Elaidin test.**

Olein on treatment with nitrous acid yields the solid '**Elaidin**'; more highly unsaturated acids remain unchanged. Consequently in this test oils yield more or less solid products in proportion to the amount of olein which they contain. The **non-drying oils** yield solid products, the **semi-drying-oils** a buttery mass separating from a liquid portion, whereas the **drying oils** remain quite liquid. The following classification is based on this test:

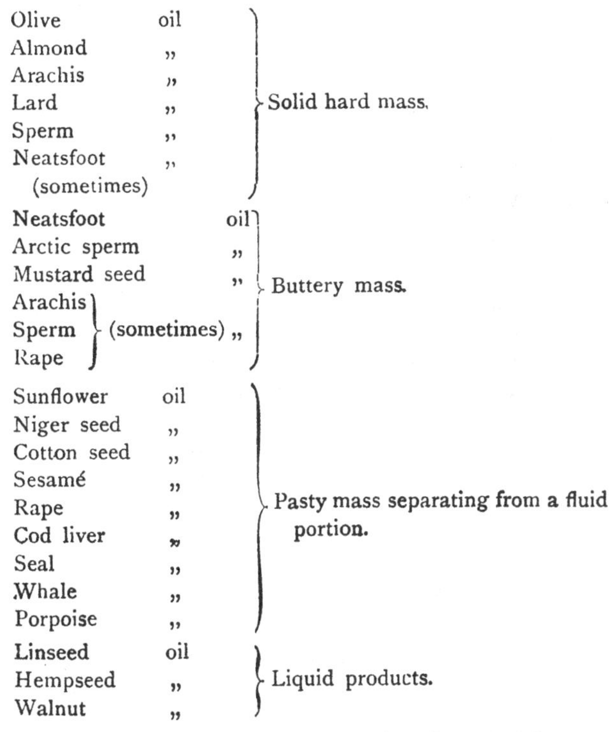

Olive	oil	Solid hard mass.
Almond	„	
Arachis	„	
Lard	„	
Sperm	„	
Neatsfoot (sometimes)	„	
Neatsfoot	oil	Buttery mass.
Arctic sperm	„	
Mustard seed	„	
Arachis, Sperm, Rape (sometimes)	„	
Sunflower	oil	Pasty mass separating from a fluid portion.
Niger seed	„	
Cotton seed	„	
Sesamé	„	
Rape	„	
Cod liver	„	
Seal	„	
Whale	„	
Porpoise	„	
Linseed	oil	Liquid products.
Hempseed	„	
Walnut	„	

[1] The results published of this test appear to show that no satisfactory method of performing it as a quantitative test has yet been found. Figures under the drying and semi-drying oils have therefore been omitted.

§ 75. SAPONIFICATION VALUE.

The number of grams of oil saponified by **one equivalent** of potash, i.e. 56·1 grams of potash, is termed the **saponification equivalent**; whilst the SAPONIFICATION VALUE is the number of MILLIGRAMS OF POTASH required to saponify ONE GRAM OF OIL,

i.e. 56,100 m. grams of KOH saponify x gram of oil (i.e. sap. equiv.)

$$\therefore \quad \frac{56{,}100}{x} \text{ m.g.} \quad " \quad " \quad 1 \ " \quad " \text{ (i.e. sap. value)},$$

$$\text{or} \quad \frac{56{,}100}{\text{sap. equiv.}} = \text{sap. val. and } \frac{56{,}100}{\text{sap. val.}} = \text{sap. equiv.}$$

Hence the saponification value *increases* as the molecular weight of the glyceride *decreases*. This is shewn in the table of the calculated saponification values of the pure triglycerides at end of book.

Free fatty acid in a fat raises its saponification value. It is therefore advisable to determine the free acid, if any, in the fat before saponification.

The presence of any considerable quantity of unsaponifiable matter, of course, proportionally reduces the figure. This factor should therefore be ascertained in conjunction with the saponification value before deductions are made as to the kind of oil or fat present.

The chief use of this figure for discriminating purposes is in the case of the

Cocoa-nut group, Porpoise and dolphin oils,
Butter fat, Rape oil class.

(*See table opposite*)

§ 76. TECHNICAL ASPECTS.

1. Soap-making

As the saponification value is a measure of the weight of potash required to combine with a given weight of oil, it indicates exactly the amount of oil and of alkali required to manufacture a given quantity of soap. The figure is in terms of potash, but a simple calculation is only necessary when producing hard soap with soda. Thus:

A ton of soft soap is required of a strength of 38 per cent. fatty acid. The saponification value of the oil is found to be 190. How much oil and potash will be required?

Assuming 95 per cent. of insoluble fatty acids are yielded from the oil, the amount of oil required will be:

$$\frac{38 \times 100}{95} \text{ per cent. of 20 cwt. or } \mathbf{8\ cwt.}$$

Assuming the potash to be (for instance) 92 per cent. purity KHO the amount required will be

$$\frac{190}{1000} \times \frac{100}{92} \times 8 \text{ or } \mathbf{1\ cwt.\ 73\ lbs.} \text{ (nearly)}.$$

A potash lye of suitable strength must be taken, and sufficient water added to give a soap of the required fatty acid.

In the case of soda, the amount of alkali required, assuming again a strength of 92 per cent. NaOH, will be obtained by multiplying the result as above by $\frac{40}{56 \cdot 1}$ (equivalent weight caustic soda, divided by the equiv. weight caustic potash).

Thus $\frac{190}{1000} \times \frac{100}{92} \times \frac{40}{56 \cdot 1} \times 8$ or **1 cwt. 20 lbs** nearly.

2. **Glycerine production.**

Assuming no free fatty acids present in an oil or fat, the theoretical yield of glycerine may be calculated from the saponification value of the oil. For this purpose, it is only necessary to multiply the **figure of the s.v. by the factor ·054664.** For since glycerol combines with 3 molecules of a fatty acid its combining weight is $\frac{1}{3}$ of the molecular weight $= \frac{92}{3}$ or $30 \cdot \dot{6}$. This is equivalent to 56·1 of Potash, therefore the factor is $\frac{30 \cdot \dot{6}}{56 \cdot 1 \times 10}$ or ·054664.

In order to know the actual yield *in practice*, the percentage of decomposition obtained must be found. From this is subtracted the original **percentage** free acid, and the final percentage obtained is multiplied by the above theoretical figure for glycerine. Thus:

Oil had 5 % free fatty acids.

After deglycerination, 88 per cent. of fatty acids were found present.

Amount of decomposition = 83 %.

Yield of glycerine $= \frac{83}{100} \times 10 \cdot 55$ or 8·75 per cent.

Small errors may be due to (1) mono- and diglycerides contained in the original oil (due to decomposition of the triglycerides); (2) mono- and diglycerides present in the fatty acids after deglycerination; (3) polyglycerides in commercial glycerine.

§ 77. ETHER INSOLUBLE BROMIDE VALUE.

On treating the unsaturated glycerides or their fatty acids with bromine, there are obtained **bromides** which differ in their solubility in solvents, especially in ether.

Thus the hexa- and octobromides from linolenic and clupanodonic acids and their glycerides, are very sparingly soluble in cold ether.

Consequently, on weighing the precipitates obtained by brominating an ethereal solution of the fat or fatty acids, an estimate is formed of the acids present which absorb 6 or 8 atoms of bromine.

The **drying** oils[1] are thus broadly distinguished from the **semi-drying** oils.

[1] It must be pointed out however that tung oil and poppy seed oil which are both classed among the drying oils yield no hexabromides.

The importance of this method of distinction will be seen when it is remembered that mixtures of drying and semi-drying with non-drying oils may be readily made which give identical iodine values, but which are readily distinguished by means of this test.

The melting point of the bromides establishes an important distinction between hexa- and octobromides, the latter as obtained from the fish oils. The hexabromides melt at about 170°—180° C. to a clear liquid, whilst the crude octobromides from fish, liver and blubber oils do not melt, but at about 200° C. begin to darken.

Marine oils can thus be distinguished from the vegetable drying oils.

From the yield of bromides the proportions of the unsaturated glycerides or fatty acids can be approximately calculated by reference to the following table.

Factors for Conversion of Insoluble Bromides into Glycerides and Acids.

Acid or Glyceride	Bromide	Factor
Clupanodonein	$C_3H_5(C_{18}H_{27}O_2Br_8)_3$	·318
Clupanodonic acid	$C_{18}H_{28}O_2Br_8$	·301
Linolenein	$C_3H_5(C_{18}H_{29}O_2Br_6)_3$	·377
Linolenic acid	$C_{18}H_{30}O_2Br_6$	·367

TECHNICAL ASPECTS.

The presence or absence of fish oils in drying oils is an important consideration in the paint and varnish industry. Generally speaking, in spite of some evidence to the contrary, it may be taken that the fish oils do not make as satisfactory paints as the vegetable drying oils. There is always a temptation to adulterate drying oils with fish oils owing to the low prices ruling for the latter.

It is therefore very necessary to be able to discover the presence of fish oil in a paint or varnish, or in an oil supplied for the manufacture of these (see chapter on drying oils).

§ 78. THE REICHERT-MEISSL VALUE.

This test is a **measure** of the **volatile acids** contained in an oil or fat. It would of course be preferable to ascertain the total fatty acids volatile in steam. Attempts in this direction have however failed owing to the fact that on repeated distillation the non-volatile fatty acids undergo decomposition yielding volatile products. As the test is therefore an **arbitrary** one the greatest care must be taken to obtain the **same conditions** in successive tests.

For this purpose definite dimensions for the apparatus and minute directions as to procedure are laid down and must be **strictly adhered to** if uniform results are to be obtained.

Most oils yield a value of below 0·5. The test is of greatest value in the examination of **butter-fat**, which gives a high figure owing to the amount of butyric acid which it contains. Porpoise and dolphin oil are also characterised by high values owing to the presence of valeric (?) acid. The cocoanut group also yield discriminating results with this test and the succeeding test.

Table of Reichert-Meissl Values.

Butter-fat	26—33	
Porpoise oil	23—40	2·5 grams of oil
Porpoise Jaw oil	48—66	2·5 grams of oil
Dolphin oil	5—6	2·5 grams of oil
Dolphin Jaw oil	66	2·5 grams of oil
Cocoanut oil	7	
Palm kernel	5—6	

TECHNICAL ASPECTS.

1. **Edible fats.** This test, and the succeeding test, are of use mainly in establishing the purity of butter fat, and in detecting admixture of animal fats and of cocoanut or palm kernel oils (q.v.).

2. **Fatty acid production.**

In the commercial production of fatty acids, the proportion of volatile fatty acids is of importance, as although the glycerides of these fatty acids yield their full amount of glycerine, most if not all of the volatile acids are of no commercial value, and further they are lost during the process of deglycerination.

It follows that a soap of somewhat different character is produced if made, not directly from (for example) cocoanut oil, but from the commercial fatty acids obtained on deglycerination of the oil.

§ **79. The Polenske Value.** This test is of use in detecting the presence of cocoanut or palm-kernel oils—especially in admixture with butter fat. It is carried out in almost an identical manner with the Reichert-Meissl Value.

The oils contain glycerides of caproic, caprylic, capric and lauric acids which pass over in the steam but are partially or wholly insoluble in water. The insoluble volatile acids are estimated by titration with alkali.

The figures obtained in the case of butter, etc. are as follows:

Fat, etc.	No. of c.c. $\frac{N}{10}$ KOH.
Butter-fat	2·3—3·3
Cocoanut oil	15—20
Palm kernel oil	10—12

§ **80. The Acetyl value.** It has already been shewn that the H of a **hydroxyl group** attached to a carbon atom is replaceable by an

acetyl radicle ($CH_3.CO$) except in the case of the $-OH$ group of the carbonyl group. This reaction forms the basis for the determination of

(1) **Alcohols** including **glycerol** (see acetin method), e.g.

$$R-OH+CH_3CO.Cl=R.O.OC.CH_3+HCl$$
$$C_3H_5(OH)_3+3CH_3CO.Cl=C_3H_5(O.OC.CH_3)_3+3HCl.$$

(2) **Monoglycerides** and **Diglycerides.**

$$C_3H_5(OH)_2(OX)+2CH_3CO.Cl=C_3H_5(O.OC.CH_3)_2(OX)+2HCl$$
$$C_3H_5(OH)(OX)_2+CH_3CO.Cl=C_3H_5(O.OC.CH_3)(OX)_2+HCl.$$

(3) **Hydroxy** or **Hydroxylated acids.**

$$C_{17}H_{32}(OH).COOH+CH_3.CO.Cl$$
$$=C_{17}H_{32}(O.OC.CH_3).CO.OH+HCl.$$
$$C_{17}H_{29}(OH)_6.CO.OH+6CH_3.CO.Cl$$
$$=C_{17}H_{29}(O.OC.CH_3)_6.CO.OH+6HCl.$$

The test is chiefly useful in recognizing and determining oils of the **castor oil** group[1].

TECHNICAL ASPECTS.

In practice the test refers only to castor oil, the technical considerations in connection with which are described on page 144

§ 81. THE ACID VALUE.

As previously stated (§ 23) oils and fats as produced in commerce, often contain small quantities of fat hydrolysing enzymes (or enzyme producing moulds) derived from the seeds or tissue. These, especially in the presence of moisture, split the glycerides, producing very varying quantities of free fatty acids.

Generally speaking, the percentage of free acids in an oil is a good guide to the grade of oil and to the care taken in its preparation. It must be noted however that oils which have been treated with caustic soda during refinement (cotton seed oil, etc.) will be almost neutral in all cases, while the enzyme is destroyed in oils which have been refined with sulphuric acid.

TECHNICAL ASPECTS

1. **Edible oils.**

More than quite a small percentage of free fatty acids present in an edible oil imparts an unpleasantly sharp flavour to the oil. This test is therefore of great importance.

2. **Soap-making and Glycerine production.**

As free fatty acids imply decomposition of the oil, the glycerine represented by the amount of free acid present is already lost. The oil is therefore of proportionally lower value for glycerine production. This test is employed to ascertain the percentage of decomposition of the oil during and after deglycerination.

[1] The distinction between the "apparent" and the "true" acetyl value is referred to in the description of the test.

3. **Lubricating.**

Free fatty acids in any amount are very objectionable in lubricating oils or fats.

Fatty acid has a very pronounced corrosive action on iron and steel, and a pitting action on most other metals, and its presence in lubricants is therefore very undesirable.

Apart from this, the free acids have often a great tendency to crystallize out and may thus cause blocking and stoppages in oil ducts.

In cylinder oils of all kinds free fatty acids are extremely objectionable, as the corrosive action is accelerated by high temperatures.

4. **Candle manufacture.**

Neutral oil (glycerides) in any quantity is objectionable in candle material for three reasons.

(1) There is loss of glycerine through the decomposition of the oil not being carried far enough.

(2) Neutral oil in the "stearine" means a lower melting point, and hence less value.

(3) On burning candles produced with such stearine, acrolein is produced, which on extinction of the candle causes an objectionable odour.

This is thus an important test for the "stearine" or fatty acid for candle material.

5. **Pharmaceutical and Medicinal.**

Oils for this purpose should contain no free fatty acid in this test, as it has an injurious action on the skin and mucous membranes.

§ 82. UNSAPONIFIABLE MATTER.

The actual amount of unsaponifiable matter in natural oils and fats is variable but the great majority contain under 2 per cent. (see § 19). Exceptions are shark, dolphin and porpoise oils (q.v.) which contain spermaceti.

As mineral oils, rosin oils and tar oils are of low value compared with the natural oils and fats, there is always an inducement to add these as adulterants. The amount if present of mineral adulterants can be accurately determined by this method, by subtracting the average figure for the pure oil from the result obtained.

For the **identification** of mineral, rosin and tar oils the unsaponifiable matter must be separately examined.

For the identification of phytosterol and cholesterol in the unsaponifiable matter see practical portion of book.

The Waxes.

As the **natural waxes** are mainly esters of monohydric alcohols, insoluble in water, they yield by this method, large percentages of unsaponifiable matter. In addition, free alcohols and hydrocarbons are often present.

TECHNICAL ASPECTS.

Considerable amounts of unsaponifiable matters may be found in oils which have been extracted by means of volatile solvents (see page 214). This is due to incomplete evaporation of the solvent, or to the employment of a too crude or imperfectly fractionated solvent.

1. **Soap-making.**

A higher percentage of unsaponifiable matter than about 2—3 per cent. renders an oil unsuitable for soap-making, especially for soft soap-making. "Shark oils" often contain unsaponifiable matter over 4 per cent.

Soap made with such oils does not "set" on cooling. "Oleine" (crude oleic acid) which is used largely in the preparation of soft soaps for wool scouring and for laundry purposes, if prepared by distillation contains notable amounts of unsaponifiable matters (chiefly hydrocarbons). The same remarks apply to distilled products of wool fat or "recovered grease." Such oleines are unsuitable for soft soap manufacture. The addition of mineral, rosin, and tar oils to oils for soap-making purposes is equally objectionable.

2. **Lubricating oils.**

Since lubricating oils normally consist of mixtures containing mineral oils, the presence of notable quantities of unsaponifiable matters in a vegetable or animal lubricating oil is mainly objectionable from the point of the diminution in value, since mineral and other unsaponifiable oils are cheaper in price.

3. **Candle industry.**

As most candles are mixtures of fatty acids with mineral waxes, this test is of importance in the determination of the latter constituent.

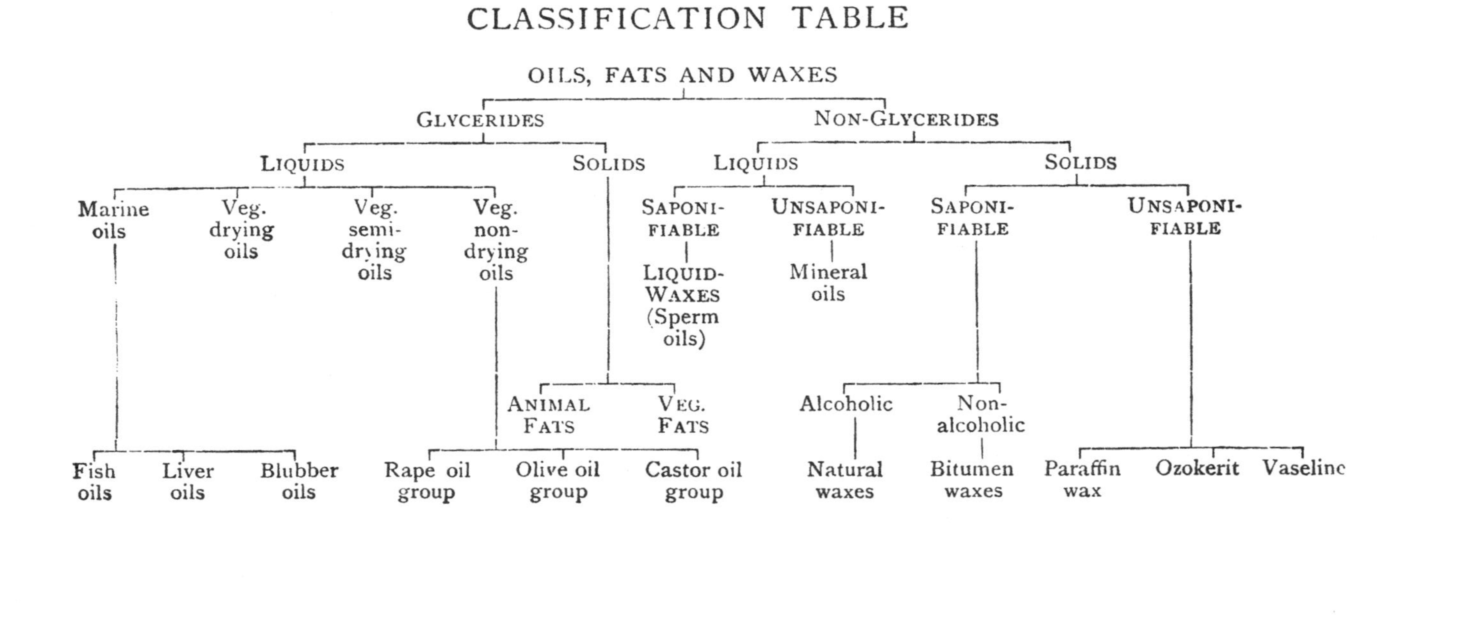
CLASSIFICATION TABLE
OILS, FATS AND WAXES
GLYCERIDES
NON-GLYCERIDES
LIQUIDS
SOLIDS
LIQUIDS
SOLIDS
Marine oils
Veg. drying oils
Veg. semi-drying oils
Veg. non-drying oils
SAPONIFIABLE
UNSAPONIFIABLE
SAPONIFIABLE
UNSAPONIFIABLE
LIQUID-WAXES (Sperm oils)
Mineral oils
ANIMAL FATS
VEG. FATS
Alcoholic
Non-alcoholic
Fish oils
Liver oils
Blubber oils
Rape oil group
Olive oil group
Castor oil group
Natural waxes
Bitumen waxes
Paraffin wax
Ozokerit
Vaseline

SECTION IV

CLASSIFICATION OF NATURAL OILS, FATS AND WAXES.

§ **83.** It has been pointed out in a previous chapter, that the fatty oils and fats are readily distinguished from the waxes since the latter are compounds of alcohols (mainly monohydric) other than **glycerol**, whilst the former are glycerides and yield glycerine on hydrolysis or saponification with alkalies.

The oils and fats may also be classified as regards their source from animals and plants. Animal oils contain the alcohol **cholesterol**, whilst oils from vegetable sources are characterized by the presence of **phytosterols**[1].

§ **84.** For further distinction between individual oils and fats, it is obvious that the character of the fatty acid radicle must form the basis of classification. In a few cases, a distinction may be made by the **molecular weight** of the fatty components of the oils.

This however applies to only a very few cases and the great majority of oils are uniformly composed of glycerides of fatty acids which are almost identical in molecular weight.

§ **85.** Another possible basis of classification is the presence of **hydroxyl** groups in the fatty acid molecule. This distinction however applies in the case of only one of the oils of commerce[2].

§ **86.** The only other distinction between the oils consists in the degree of unsaturation of the component fatty acids.

There are here two aspects to consider, viz.:

1. The degree of unsaturation of the individual acids.
2. The amount of unsaturation of the total acids present expressed in terms of their iodine absorption.

Generally speaking, those oils which contain the more highly unsaturated substances, give also the higher iodine absorption figures.

There is however one important exception. The **fish oils** of commerce —more correctly termed the **marine oil group**[3]—are characterized by the presence of the most highly unsaturated glycerides occurring naturally, but have not commonly such high iodine absorption values

[1] Sisosterol, etc. [2] Castor oil.
[3] Since the blubber oils are from marine mammals.

as these vegetable oils which contain glycerides of a lower degree of unsaturation.

Further, certain oils, sufficiently high in iodine absorption values to be placed with oils such as **linseed**, contain actually no acids of like degree of unsaturation.

As however these oils, notably tung and poppy seed, approximate closely to linseed oil in their properties, it is convenient to group them in the same class.

§ **87.** It therefore follows that the most rational classification, and the one adopted here, is:

1. To place the oils in groups according to the **character** of the most highly unsaturated fats they contain.

2. To classify each individual member of the group according to its total unsaturation value, as determined by its iodine absorption.

There are thus **four** main groups, as follows:

A. Oils containing glycerides of fatty acids which absorb **8 atoms** of bromine (e.g. Clupanodonic acid), i.e. Marine oils.

B. Oils containing glycerides of fatty acids which absorb **6 atoms**[1] of bromine (e.g. Linolenic acid).

These are the vegetable drying oils.

C. Oils containing notable proportions of glycerides of acids which absorb **4 atoms** of bromine (e.g. Linolic acid).

These are the semi-drying oils.

D. Oils containing little or no glycerides of acids more unsaturated than **Oleic**, absorbing **2 atoms** of bromine only. (This group includes the fats of commerce. There are however fats known which belong to group C[2].)

The natural waxes may be distinguished as:

1. Solid.

2. Liquid.

Also as to their source, i.e. animal and vegetable. This distinction is however not ascertainable by chemical tests, as in the case of oils and fats.

§ **88.** With regard to the **Analytical data**, only those have been included which yield definite information. Each test is distinguished by a definite number throughout, as follows:

(**1**) Specific gravity at 15·5° C.
(**2**) Melting point ° C.
(**3**) Solidifying point fatty acids ° C.
(**4**) Refractive index (Z. B.)[3] at 40° C.

[1] With exceptions as before stated.

[2] These are of no practical interest at present.

[3] Zeiss Butyrorefractometer. Nearly all the determinations are made on this instrument.

(5) Viscosity.
(6) Solubility ("true Valenta") and "Turbidity temperature."
(7) Polarimeter.
(8) Iodine value.
  a. Bromine Thermal test.
  b. Maumené test.
  c. Livache test.
(9) Elaidin test.
(10) Saponification value.
(11) Insoluble Bromide value.
(12) Reichert-Meissl value.
  a. Polenske value.
(13) Acetyl value.
(14) Acid value.
(15) Unsaponifiable matter, per cent.

Those of the above data which are of importance and yield definite information in connection with the various classes of oils, fats and waxes are indicated in the introduction to each of the groups and classes in question.

The following numbers have importance only in the case of a **single** class of oils:

(5) Lubricating oils.
(9) Non-drying oils.

Others are of value only in connection with **particular** oils, as:

(7) and (13). Castor and a few rare oils.
(12) and (12*a*). Coconut group and butter fat.

In all cases the figures which are of importance are indicated under the description of each individual oil, fat and wax, *even where these have not as yet been ascertained*[1] with certainty. In this case, a **blank space** is left opposite the number of the test thus indicating where further work is desirable.

Extreme values have, from time to time, been recorded by various observers, and these, being outside the usual range, are given separately under each oil, fat and wax, and the authorities indicated in each case.

[1] In many cases the authors have these determinations in hand, but were unable to complete them before going to press.

THE MARINE OILS

Group 1.	FISH OILS.	Typical oil,	Menhaden.
Group 2.	LIVER OILS.	,, ,,	Cod Liver oil.
Group 3.	BLUBBER OILS.	,, ,,	Whale oil.

A. GLYCERIDES

(I) LIQUIDS

This group comprises all the natural oils.

CHAPTER VI

CLASS I. THE MARINE OILS

Oils containing glycerides of fatty acids absorbing 8 *atoms of bromine.*

§ **89.** These oils are all similar in their physical characteristics, possessing, especially in the crude condition, a pronounced and more or less disagreeable 'fishy' odour and acrid flavour.

Chemically they contain varying proportions of the highly unsaturated acid termed by Tsujimoto, **clupanodonic acid**, absorbing 8 atoms of bromine per molecule, giving the octobromide $C_{18}H_{28}O_2Br_8$. Similarly on hydrogenation in presence of a catalyst, this acid is converted into stearic acid with the absorption of 8 atoms of hydrogen. Analytically these oils are distinguished from the vegetable drying oils—which they resemble in their high iodine absorption values—by the behaviour of the octobromides obtained on brominating the mixed fatty acids. (See Insoluble Bromide Values, p. 86.)

These bromides on heating do not melt but begin to darken and decompose at a temperature of 200° C.

A further distinguishing point is that these oils, in the Livache Test (q.v.), though absorbing large amounts of oxygen, do not dry to an **elastic** skin.

Possible exceptions are the two oils, **porpoise and dolphin**, which require further investigation, especially the former.

§ **90.** The marine oils form three subdivisions, according to their origin, viz.:

1. **Fish oils** obtained from the bodies of various fish.
2. **Liver oils** derived from the livers only of a certain class of fish.

3. **Blubber oils** contained in the oil-tissue of marine mammals[1].

Liver Oils are distinguished from fish and blubber oils by the presence of substances derived from the liver cells, of which cholesterol is the most important. These give a colour reaction with sulphuric acid described later (see practical part).

The **blubber oils** differ enormously in their composition. Some resemble the fish oils very closely, others approximate to the **liquid waxes** in their large content of non-glycerides and are also remarkable in containing large amounts of glycerides of fatty acids of low molecular weight.

Notes on Analytical Data.

§ **91.** The most important physical and chemical data are:

(1) Specific gravity.
(3) Solidifying point of fatty acids.
(4) Refractive index.
(6) Solubility ("true Valenta")[2].
(8) Iodine value, and its variants, bromine thermal test, etc.
(10) Saponification value.
(11) Insoluble bromide value.
(14) Acid value.
(15) Unsaponifiable matter.

The **melting point** (2) is of little value for discriminative purposes owing to the very varying amounts of 'stearine' contained in these oils.

As, owing to their pronounced gumming properties when exposed to air, these oils are quite unsuitable for lubricants[3], the estimation of the **viscosity** (5) has no significance.

The slight optical activity which marine oils exhibit is due, not to the glycerides, but to the varying amounts of the unsaponifiable constituents. Hence the **polarimeter** (7) gives no definite information.

The **elaidin** reaction (9) is negative in all cases.

The amount of volatile fatty acids is small and uncharacteristic[4], hence the **Reichert-Meissl** (12) and **Polenske** (12*a*) tests are unimportant.

Apart from the **iodine value** (8) which is somewhat variable in different samples of marine oils, by far the most important discriminative test is the insoluble bromide value of fatty acids (11) giving a measure of the percentage present of **clupanodonic acid,** the characteristic acid of the class.

[1] With the important exception of the **Sperm whale** (see 'Sperm oil').
[2] See footnote on page 75.
[3] Note however exceptions in case of blown whale oil, and the oils from porpoise and dolphin.
[4] Exceptions are dolphin and porpoise oils.

Chemical Composition.

Very imperfectly known. **Oleic acid** is not proved to occur. The 'stearine' is probably **palmitic acid.** Clupanodonic acid is present in varying proportions.

TECHNICAL.

The chief efforts towards refinement of marine oils have in general been directed to the elimination of the unpleasant fishy odour which they all possess in a greater or less degree. *Tsujimoto* attributes the odour to the clupanodonic acid which they contain, and this opinion would appear to receive confirmation from the fact that hydrogenated oils are practically odourless. Against this view may be cited the fact that **linolenic** acid was not at first obtained odourless, and the odour of clupanodonic acid may therefore conceivably be due to impurities, and further, that it is not until the hydrogenation of fish oil is almost complete, that the odour entirely disappears. Many methods have been proposed to remove the odour without altering the composition of the glycerides—some with a certain amount of success.

Group 1. Fish oils.

§ 92. TYPICAL OIL. MENHADEN.

The fish oils exist in most fish to an equal extent in almost all parts ot the tissues.

There are four oils of commercial importance, viz. :

Menhaden oil
Japan fish oil
Herring oil
Salmon oil.

The method of obtaining the oil is by boiling or steaming the fish whole or sliced up, either at atmospheric or higher pressures. The partially exhausted mass is then usually pressed, yielding a further quantity of oil. In the case of salmon oil, this procedure is only used when the fish has become putrid. Normally it is a product of the edible salmon in the tinning industry.

§ 93. TYPICAL OIL. MENHADEN OIL.

I. General and Analytical.

Character.

COLOUR varies from pale yellow to brown.

ODOUR faintly to strongly 'fishy.'

"STEARINE" variable in amount.

Physical and Chemical Data.

	Test	Average figure	Usual variations	Extremes recorded
(1)	Specific gravity at 15·5° C.	**0·931**	·929—·933	·927[1]—·936[1]
(3)	Solidifying point ° C. fatty acids			
(4)	Refractive index 40° C. (Z. B.)	**72°**		
(6)	Solubility. True Valenta $\frac{V \times 10}{80}$ Alcohol reagent			
(8)	Iodine value	**160**	150—170	147·9[2]—192·9[1]
	(*a*) Bromine thermal test ° C.	**29°**		
	(*b*) Maumené ° C.	**125°**	123—128	
	(*c*) Livache (O absorption)			
(10)	Saponification value	**193**	190—195	188·7[3]—
(11)	Insoluble bromide value fatty acids			63·1[4]—64·5[4]
(14)	Acid value	**7**	5—8	3[5]—12[5] very low grades higher
(15)	Unsaponifiable (per cent.)	about **1%**	·8—1·2	·6[6]—1·6[7]

Remarks. The tests which are not recorded have no significance in the case of this oil. The most important are (8) and (11) for discriminative purposes, and (14) and (15) for determination of grade of oil.

Chemical Composition.

Not fully known. From (11) the calculated percentage of clupanodonic acid is about 19—20.

Adulteration.

Prices usually lower than vegetable drying oils, but dearer than **mineral** and **rosin** oils, with which it may be adulterated. When present these are readily detected by the lowered saponification value and increased percentage unsaponifiable matter. For detection of rosin oil the *Liebermann-Storch* test is useful.

Special tests. Hoppenstedt's colour reaction, see practical part.

II. Technical.

Source. The body of the American fish *Alosa Menhaden*, resembling a herring.

Content of oil. Varies, but not usually over 15 per cent.

[1] Lewkowitsch. [2] Archbutt. [3] Bull.
[4] Atlas (E. and M. method in author's laboratory).
[5] Alsop. [6] Fahrion. [7] Thompson.

Industry.

Chiefly at New Jersey and Atlantic coastline N. America. Season, May to November. Two methods are used for obtaining oil:

(1) Boiling in large pans with false bottoms. On settling, oil rises to surface and is skimmed off.

(2) By a continuous process of steaming, and subsequent settlement.

The residual 'fish scrap' is pressed in hydraulic presses, and a further yield of inferior oil obtained.

The pressed cake or 'chum' is dried, and forms a valuable manure.

The production amounts to about half a million tons per annum.

Refinement.

There are three qualities of crude oil, (1) prime crude, (2) brown-strained, (3) light strained. The lighter qualities are bleached by means of Fuller's Earth. The 'strained' oils have been filter pressed.

Properties and Uses.

The oil has a variety of uses among which the chief are:

Currying leather.
Adulterating paint oils.
Soap-making.
Adulterating cod liver oil.
Manufacture of linoleum.
Tempering of steel.

§ 94. "JAPAN FISH OIL."

The oil known in commerce by this title may be either **sardine oil** or **herring oil**, or mixtures of these with lesser known fish oils. The figures for each of these oils are given, together with those obtained for the commercial oil.

I. General and Analytical.

Character.

Yellow to brown, with varying degrees of 'fishiness' in odour and flavour, and of consistence, depending on grade. Stearine is often absent at 15° C.

Physical and Chemical Data.

	Test	Japanese Sardines	Japanese Herring	Commercial oil variations
(1)	Specific gravity 15·5° C.	·932—·934	·918—·925	916—934
(2)	Solidifying point fatty acids °C.	35°—36°	30—31	28—
(3)	Refractive index 40° C. (Z. B.)			56—61
(6)	Solubility. True Valenta $\frac{V \times 10}{80}$			
	Alcohol reagent			58—68
(8)	Iodine value	181—187		121—171
	(a) Bromine thermal test			
	(b) Maumené			
	(c) Livache			
(10)	Saponification value	195—196	186—190	189·8—192
(11)	Insoluble bromide value	44—47	13—21	
(14)	Acid value			10—40
(15)	Unsaponifiable matter (per cent.)			·5—1·5

Special tests. None.

Remarks.

The cruder oils contain large amounts of albuminous substances which give very close emulsions on boiling the oil with water or weak acid.

Chemical Composition.

Not fully known. The "stearine" is probably **palmitin**. Clupanodonic acid is proportional to the insoluble bromides found, being about 0·3 of this amount. It varies thus:

Japanese sardine	13·5 to 14 per cent.
Japanese herring	3·5 to 6·5 „ „
"Japan fish" (commercial)	6·5 to 7 „ „

Adulteration.

Too cheap to adulterate except with mineral and rosin oils, easily detected by determining the unsaponifiable matter. For detecting rosin oil, see *Liebermann-Storch* reaction.

II. Technical.

Source.

The Japanese sardine, *Clupanodon melanosticta*, the Japanese herring, *Clupea pallasi*, and other less known fish oils.

Method of obtaining oil.

In general, as for Menhaden oil, but a continuous process is not yet in operation.

(1) The fish are chopped up, boiled with water, and the oil skimmed off.

(2) When labour is scarce, the fish are allowed to rot in heaps, when the oil flows out. A further yield is obtained on applying pressure.

The first method yields of course the better quality oil.

Refinement.

Chiefly at Yesso and Yokohama. It is quite primitive, and merely consists in heating up the oil and allowing to settle, when some of the albuminous impurities and water subside.

Uses.

In Japan, for currying leather, and as a paint oil.

In Europe, chiefly for 'Fish' soft soap. It has been successfully 'hydrogenated' (hardened) by means of hydrogen and a catalyst, and yields a hard, almost white and odourless fat.

§ 95. Herring Oil.

In addition to the Japanese herring oil, described under 'Japan Fish Oil,' an increasing amount of North Sea herring oil (from the species *Clupea harengus*) is being produced and shipped from Norway. The analytical figures of this oil do not differ materially from those of the Japanese variety. The oil occurs in commerce in varying conditions of crudity, with widely differing characteristics as regards colour and odour.

§ 96. Salmon Oil.

This oil is normally a bye product of the salmon tinning industry of British Columbia, and has a not unpleasant odour and flavour. The sample examined by the authors however (from a consignment of several hundred tons) was of extremely rank odour and reddish in colour. It was undoubtedly the product of putrid fish. It gave the following characteristics:

(1)	Specific gravity at 15·5° C.	·9242
(3)	Solidifying point fatty acids.	24·5°
(4)	Refractive index, 40° C. (Z.B.)	68·5
(8)	Iodine value.	168
(10)	Saponification value.	188·5
(14)	Acid value.	45·5

The oil is used for soap-making and for treating leather.

§ 97. Group 2. Liver Oils.

TYPICAL OIL—COD LIVER.

The livers of certain fish are very rich in oil. Of these, the principal and best known is the **Cod**, but oil is also obtained from the livers of the Shark, Ling, Haddock, Skate and others.

The method of extraction described under Cod Liver Oil applies to all cases.

The oils are characterised by containing minute quantities of substances which give colour reactions and are derived from the liver cells. Of these **colour reactions**, many have been proposed, but the only one really reliable is the sulphuric acid test.

§ 98. COD LIVER OIL.

I. General and Analytical.

Character.

Water white to dark brown in COLOUR.

ODOUR faint to nauseous. TASTE very mild to acrid.

"STEARINE" varies in amount according to the temperature of extraction, and the amount of "racking" (*vide infra*).

Chemical Composition.

Further research is necessary to establish the composition of cod liver oil. An acid exists which gives an OCTOBROMIDE, but there is doubt whether this is Clupanodonic acid or the Therapic acid of Heyerdahl. Among the solid acids were found Myristic, Palmitic, Stearic and Erucic acids[1].

The CRUDE OIL contains organic bases, among which are amines and ptomaines[2]. A small amount of IODINE[3] (·0002—·03 per cent.) is stated to occur in the oil.

Physical and Chemical Data.

	Test	Average figure	Usual variations	Extremes recorded
(1)	Specific gravity 15·5° C.	**·925**	·923—·930	·922[4]—·941[5]
(3)	Solidifying point fatty acids	(according to racking)	13°—24°	—
(4)	Refractive index 40° C. (Z. B.)	**70**	68°—71°	66°[6]—
(6)	Solubility. True Valenta $\frac{V \times 10}{80}$ Alcohol reagent			
(8)	Iodine value	free stearine **170**	160—175	—198[6]
		with stearine **140**	—	135[6]—
	(*a*) Bromine thermal test	**28**		
	(*b*) Maumené		113—115	100[7]—116[8]
	(*c*) Livache			
(10)	Saponification value	**185**	182—189	168[5]—193[9]
(11)	Insoluble bromide value fatty acids	(according to stearine present)	28—30	18[10]—35[5]
(14)	Acid value from practically neutral to very high acidity [in crude oils].			
(15)	Unsaponifiable matter	**1·0**	0·5—1·5	0·5[11]—4·6[12]

[1] *Bull. Berichte* 1906, 3570.

[2] Gautier and Morgues, *Comptes Rend.*, 1888 [107], 254, 626, 740.

[3] Cerdeiras (*Anal. Fis. Quim.* 1915, **13**, 4391) claims to have isolated from Cod Liver Oil the glyceride of tetrachlorotetraiodotherapic acid

$$C_3H_5(C_{17}H_{25}O_2Cl_4I_4)_3.$$

[4] Kremel. [5] Bull. [6] Lewkowitsch. [7] Henseval. [8] Baynes.
[9] Sage. [10] Hehner. [11] Fahrion. [12] Parry.

Adulteration.

FISH OILS.

For detection of admixed Menhaden oil see page 99.

SHARK LIVER OIL is often added to Cod and may generally be detected by the increased amount of UNSAPONIFIABLE MATTER. The Insoluble Bromide Value is also lower.

VEGETABLE OILS give lower insoluble Bromide values (except Linseed), and are detected by the Phytosteryl acetate test.

MINERAL OILS are detected by increase in unsaponifiable matter and are identified by examination of the same.

II. Technical.

Source.

From the liver of the cod, *Gadus morrhua*. "Coast cod oil" is from the livers of other fish including the Shark, Hake, Ling, Haddock, etc.

Industry.

The cod-fishing industry is carried on mainly off the coast of Newfoundland and in the Lofoten Islands (Norway). Of late years a large amount of oil has been produced in Japan where there are extensive cod-fishing grounds.

Formerly sailing boats were used, and the livers were cut out of the fish until a sufficient number had been collected before being landed and the oil recovered. In consequence the livers on arrival at the stations were in a more or less putrid condition and yielded a dark crude oil.

A certain amount of oil exudes from the fresh livers in the cold, and this was formerly the only quality fit for medical use.

As steam trawlers are now employed it is possible to land the fish alive. The fresh livers are then heated in steam-jacketed vessels, when the oil is freed from the liver-cells and exudes forming the MEDICINAL OIL. A further heating at a higher temperature yields a LIGHT BROWN OIL and a third grade is obtained by prolonged boiling of the livers with water. This last, together with any oil recovered from putrid livers, forms the "Brown Oil" grade.

The following qualities are therefore distinguished:

Raw Medicinal Oil.
Pale Oil.
Light Brown Oil.
Brown Oil.

According to the temperature employed these oils will contain more or less "stearine." This is often removed by settlement ("racking") and is sold as "fish tallow."

Uses. Best qualities for medicinal purposes.

The beneficial effect is considered by Lewkowitsch to be due to the character of the unsaturated glyceride, which he considers more easily assimilated than those of other oils.

Currying trade. Soap-making.

§ 99. Shark Liver Oil.

Character. As for Cod Liver Oil [from which it is scarcely distinguishable].

Remarks. The high unsaponifiable and low saponification values are very noteworthy. Two distinct varieties of oil apparently exist, with possible intermixtures.

Chemical Composition. Imperfectly known.

Special Tests. None.

Physical and Chemical Data.

		(A)	(B)[1]
(1)	Specific gravity 15·5° C.	·915 to ·917	0·8666
(4)	Refractive index 40° C.		
(8)	Iodine value	114 to 155	358
(10)	Saponification value	146 to 190	22·5
(11)	Insoluble bromide value		76·5
(14)	Acid value	Very variable	0·84
(15)	Unsaponifiable	0·7 to 17 per cent.	89·1

Adulteration. Not specially liable to adulteration owing to low market value. Mineral and Rosin oils might and would be detected by the Iodine Value of the unsaponifiable matter[2].

Technical.

Obtained from the **shark** (*Scymnus borealis*) during cod fishing and in Japanese waters. The oil with the high unsaponifiable matter is from *Centrophorus granulosus* and *Scymnus lichia* of the family *Spinacidae*. For details of extraction see Cod Liver Oil.

Uses. As adulterant of Cod Liver Oil. Manufacture of tarpaulins.
Soap-making (oils with high unsaponifiable matter unsuitable)[3].

§ 100. Group 3. Blubber Oils.

Typical Oil—Whale.

These oils, although obtained from mammals, resemble very closely the **fish oils** in their physical characteristics.

Whale and seal oils are also closely allied chemically with the fish oils. The oils obtained from the porpoise and dolphin are totally different in chemical composition, containing large amounts of the non-glyceride **spermaceti**, and being thus intermediate between whale and seal

[1] Chapman, *Analyst* 1917, **42**, 161. These extraordinary figures are due to the preponderance of a hydrocarbon named "Spinacene" in the oil. It is very highly unsaturated, absorbing 12 atoms of bromine: formula $C_{30}H_{50}$ (bromide $C_{30}H_{50}Br_{12}$).

[2] Iodine value of unsaponifiable matter: 376·2.

[3] New uses for spinacene-containing oils will probably be found.

oils on the one hand and sperm oil on the other, which is a true wax—a non-glyceride.

Another curious point is the difference in composition between the oil derived from the head and that from the body of the dolphin and porpoise.

Further, these oils are rich in the glycerides of volatile fatty acids, probably butyric and caproic, and thus yield very high Reichert values.

The method of extraction of all the blubber oils resembles that employed for whale oil, which is described in detail.

§ 101. SEAL OIL.

I. General and Analytical.

Character.

Water-white to brown in COLOUR. ODOUR faint to strong "fishy." "STEARINE" very variable.

Physical and Chemical Data.

	Test	Average figure	Usual variations	Extremes recorded
(1)	Specific gravity at 15·5° C.	·925	·9244—·9261	—·9336[1]
(3)	Solidifying point fatty acids	15°	14°—16°	13°[1]—17°[1]
(4)	Refractive index 40° C. (Z. B.)		64—65	
(6)	Solubility. True Valenta $\frac{V \times 10}{80}$ Alcohol reagent			
(8)	Iodine value	140	130—147	127[2]—193[1]
	(a) Bromine thermal test	25		
	(b) Maumené		90—94	
	(c) Livache			
(10)	Saponification value	191	189—193	178[2]—196[3]
(11)	Insoluble bromide value fatty acids	20		
(14)	Acid value dependent on grade 0·2—20 per cent.			
(15)	Unsaponifiable matter per cent.		0·4—1	

Chemical Composition. Imperfectly known. The solid acid is probably Palmitic Acid and the oil would appear to contain 5—6% of Clupanodonic Acid in the mixed fatty acids.

Adulteration.

Mineral and Rosin Oils—high unsaponifiable.
Other Marine Oils—difficult to detect.

II. Technical.

Obtained from the blubber of seals (*Phoca*) of which there are several varieties.

[1] Schneider and Blumenfeld. [2] Kremel. [3] Stoddart.

The practice of obtaining the oil varies. Sometimes it is obtained on the vessel, from the fresh blubber, yielding a pale, sweet oil; in other cases the blubber is taken to land and being more or less putrid yields lower qualities. The method of rendering follows closely that of **whale oil** (which see).

Four grades are usually recognized, viz.

Water White.	Yellow Seal.
Straw Seal.	Brown Seal.

Uses.

For Burning (in old fashioned lighthouses and on board ships). For adulterating Cod Liver Oil, and for Soap-making and production of stearine by hydrogenation.

§ 102. TYPICAL OIL. WHALE OIL.

I. General and Analytical.

Character.

Water white to almost black in COLOUR.

Nearly odourless to extremely rank and "fishy."

CONSISTENCE varies according to stearine present.

The lowest grades are semi-solid on account of albuminous impurities.

Physical and Chemical data.

	Test	Average figure	Usual variations	Extremes recorded
(1)	Specific gravity at 15·5° C.	**·922**	·917—·927	
(3)	Solidifying point fatty acids ° C.	**23°**		
(4)	Refractive index 40° C. (Z. B.)	**58**	56—59	
(6)	Solubility. True Valenta $\frac{V \times 10}{80}$ Alcohol reagent			
(8)	Iodine value	**115**	110—125	103[1]—136[1]
	(*a*) Bromine thermal test	**21**		
	(*b*) Maumené		90—92	
	(*c*) Livache			
(10)	Saponification value	**188**	187—189	178[1]—194[2]
(11)	Insoluble bromide value fatty acid	**27**	22—30	12[3]—31[4]
(14)	Acid value	according to grade ½ °/。 to almost 100 °/。		
(15)	Unsaponifiable matter	varies from 0·5 to 3·3 per cent. (The latter figure is exceptional. Most oils do not exceed 2 per cent.)		

[1] Bull. [2] Schweitzer. [3] Walker and Warburton.
[4] Atlas (in author's laboratory).

Chemical Composition.

About 8·5 per cent. of Clupanodonic acid is present in the mixed fatty acids. The solid fat appears to consist mainly of a glyceride of Palmitic acid. Further than this the composition is at present not established.

Adulteration.

Rosin and Mineral oil would be indicated by a high percentage of unsaponifiable matter. Adulteration with fish oils would be difficult to detect unless fair quantities were present. It would then be shown by an increased Iodine value (in most cases) and an increased yield of ether-insoluble bromides. Adulteration with **Shark liver oil** has been practised, and is indicated by an unsaponifiable of over 3 per cent., in the absence of Mineral or Rosin Oils.

II. Technical.

Source.

The oil is obtained from various species of the genus *Balaena*.

Industry.

The whaling industry is chiefly carried on in Norway, the Lofoden Islands, Iceland and the United States. Stations are also established in Japan and South Africa. Some of the blubber is still treated on board ship. The yield is very variable according to the species of animal. The "Right" whales yield 18 to 180 barrels per animal, whereas the "Humpback" yield only half this quantity and other species scarcely any.

At the "trying" stations where the whales are brought, the blubber is severed from the flesh, chopped fine in machines, heated, and boiled with steam in open pans. After the boiling process, the blubber which still contains oil, is further rendered with steam in closed vessels, "digesters," at pressures of 40—50 lbs. per sq. inch. This yields darker and cruder oil. The bones are similarly treated.

The following grades are produced:

No. 0 Whale Oil.		Runs off blubber in the cold. Water white to pale yellow; "sweet," i.e. faint odour.
No. 1	,,	First boiling. Pale yellow; slightly stronger odour.
No. 2	,,	Produced in digester. Light brown colour; strong smell.
No. 3	,,	Bones digested. Dark brown; pronounced "fishy" smell.

No. 4 Whale Oil. Blubber from putrid carcase.
Brown to black ; nauseous smell.

The residue forms "whale guano," a valuable manure. If *perfectly fresh* it is suitable for cattle feeding.

Production.

Total production estimated at about 16,000 tons per annum. About $\frac{1}{3}$ of this is produced in Norway, about $\frac{1}{4}$ in the United States and the remainder from the other countries.

Uses.

Hydrogenation for stearines of all grades.

Soap-making.

Burning.

Leather Dressing.

The "Blown Oil" is used as a lubricant alone and dissolved in mineral oils (forming "compound" lubricating oils).

§ 103. Dolphin and Porpoise Oils.

Both of these marine mammals yield oil from two distinct sources, viz. the **body blubber**, and the oily tissue of the head and jaw. The body oil yields varying amounts of **solid spermaceti wax**, while the 'jaw oil' is extremely rich in glycerides of volatile fatty acids (of which the body oil also contains a proportion). Thus the Reichert-Meissl Test gives with the jaw oils a figure of about 60, and the proportion of insoluble fatty acids, together with the unsaponifiable matter is only about 66. The volatile fatty acid present has been thought to be **isovaleric,** but may be a mixture of butyric and caproic acids.

The oils are somewhat rare, and are used for lubricating fine machinery.

VEGETABLE DRYING OILS

Typical oil, Linseed

CHAPTER VII

CLASS II. VEGETABLE DRYING OILS

Oils containing in general glycerides of fatty acids which absorb six atoms of Bromine.

TYPICAL OIL. LINSEED.

§ **104.** These oils are named from their property of forming a solid elastic substance when exposed to the air in thin layers. This "drying" power decreases as the Iodine absorption value of the oil diminishes, i.e. it is in ratio to the total amount of unsaturated fatty acids present, and not to the percentage of linolenic acid in the composition.

Thus Tung oil and Poppy-seed oil, which appear to contain little or no linolenic acid (or other acid absorbing 8 atoms of bromine) are both good drying oils owing to the large total quantity of unsaturated acids they contain.

They find their chief use in commerce as the vehicles for pigments in paints and varnishes, and to a lesser extent for soft soap-making.

The Iodine Values of these oils range from about **200** to **120.**

§ 105. Notes on the Analytical Data.

The important physical and chemical data are:

(**1**) Specific gravity.
(**3**) Solidifying point fatty acids.
(**4**) Refractive Index.
(**6**) Solubility ("truc Valenta")[1].
(**8**) Iodine Value, and its variants.
(**10**) Saponification value.
(**11**) Insoluble Bromide value.
(**14**) Acid value.
(**15**) Unsaponifiable.

Of these (**1**), (**4**) and (**8**) are most generally useful, and (**11**) of greatest discriminative value.

The **melting-point** (**2**) is in practically every case below normal indoor temperatures; and is unimportant, except in special cases where freezing of an oil might occur.

For remarks on the Viscosity (**5**), Polarimeter (**7**), Elaidin test (**9**), and Reichert-Meissl value (**12**), see under Marine Oils (page 96).

[1] See footnote, page 75.

§ 106. Perilla Oil.

Character.

Resembles Linseed oil in odour and flavour.

Physical and Chemical Data.

(1)	Specific gravity 15·5° C.	·9305[1]
(3)	Solidifying point fatty acids ° C.	
(4)	Refractive index 40° C.	1·4753[1]
(6)	Solubility. True Valenta	35·5°[2]
	$\frac{V \times 10}{80}$	4·4[2]
	Alcohol reagent	60·0°[2]
(8)	Iodine value	206[3]
	(*a*) Bromine thermal test	
	(*b*) Maumené	124[1]
	(*c*) Livache	
(11)	Insoluble bromide value fatty acids	64[4]
(14)	Acid value	

Remarks.

It is characterized by its very high **Iodine value**, which is the greatest of all known oils.

Technical.

Source.

The nuts of the Perilla Ocimoides, an annual plant, occurring in China, Japan and E. Indies.

Content of oil. About 36 per cent.

General.

The oil is obtained by crude native methods and is used in Manchuria largely for edible purposes. In Japan it is employed as an adulterant of **lacquer.** Curiously enough, in view of its larger content of unsaturated glycerides, the oil is inferior as a drier to Linseed.

§ 107. Typical Oil. LINSEED OIL.

I. General and Analytical.

Character.

Colour. Pale yellow to brown.

Odour. Faint, but distinctive. Easily recognized on boiling with water.

Stearine. None separates at ordinary temperatures.

Remarks.

The Iodine Value is the highest in the case of the **Baltic Oil,** which is from the purest seed. Other grades of oil containing

[1] Rosenthal. [2] Fryer and Weston. [3] Wijs.
[4] Eibner and Muggenthaler.

admixtures of other oil-seeds, do not give a maximum Iodine value above the following figures:

Black Sea 182. East Indian 187. River Plate 194.

Physical and Chemical Data.

	Test	Average figure	Usual variations	Extremes recorded
(1)	Specific gravity at 15·5° C.	**0·934**	·932—·937	·930[1]—·941[2]
(3)	Solidifying point fatty acids	**19·5°**	19°—20·5°	—
(4)	Refractive index 40° C. (Z. B.)	**73**	72—74	—
(6)	Solubility. True Valenta	47·0°[3]		
	$\frac{V \times 10}{80}$	5·9[3]		
	Alcohol reagent	62·3°[3]		
(8)	Iodine value	varies according to origin (see note)	175—200	—205[4]
	(*a*) Bromine thermal test	**31**	30·5°—32°	
	(*b*) Maumené	**115**	112—126	103[5]—145[6]
	(*c*) Livache			
(10)	Saponification lue	**190**	189—194	
(11)	Insoluble bromide value fatty acids	**35**	30—40	**50—58**[7] 29[8]
(14)	Acid value	varies up to 10, but usually not above		
(15)	Unsaponifiable matter	varies from 0·5 to 2 per cent.		

The Iodine Value (**8**) is the most important test, giving the best guide as to the purity of the oil: next in importance are the Insoluble Bromide Value (**11**) and the Livache test (**8** *c*), the latter yielding information as the nature of the dried film of oil.

Special Tests.

None.

Chemical Composition.

This is not fully established. Calculated on the amount ot Insoluble Bromides yielded by the method of *Eibner and Muggenthaler*, the proportion present of Linolenic acid would be 21 per cent. The solid acids appear to be Myristic and Palmitic acids in about equal proportions; total 7 to 10 per cent. The percentage of Linolic and Oleic acids is doubtful.

[1] Gill. [2] Lewkowitsch. [3] Fryer and Weston.
[4] Thompson and Dunlop. [5] Maumené. [6] Archbutt.
[7] Eibner and Muggenthaler. [8] Walker and Warburton.

Adulteration.

The following oils, being usually below Linseed oil in value, may be used at such times as adulterants:

MINERAL AND ROSIN OILS.

Detected by increased percentage of unsaponifiable matter. [Shark oil and Sperm "oil" will also raise this figure.]

Rosin oil is readily recognised by its optical activity.

ROSIN

Liebermann-Storch test. Estimated by TWITCHELL method (see practical part).

Adulteration with MARINE OILS is more difficult of detection.

The following two methods are reliable:

(1) MELTING POINT OF INSOLUBLE BROMIDES. The Insoluble Bromides prepared from Linseed fatty acids melt at 175°—180°. Those from marine fatty acids darken at 200° but do not melt. In the presence of over 10 per cent. of Marine oils, the bromides do not melt to a clear liquid below 180°, and they darken distinctly.

(2) MELTING POINT OF STEROL ACETATES. Obtained as described later on, the crystals of Phytosteryl acetate from pure Linseed oil melt at 128°—129°. Marine oils lower the melting point. Marine oils may also usually be detected by the characteristic fishy odour yielded on acidifying the soap solution obtained by saponifying the oil. [This is best done with dilute sulphuric acid.] The volatile odorous products of the Marine oils are in this manner rendered more distinct.

As Linseed is frequently one of the cheapest of the vegetable oils, adulteration with other vegetable oils is only likely when these happen to be lower in price.

Since all oils (with the exception of PERILLA) absorb less Iodine than Linseed, adulteration will be indicated by a lowered Iodine value. Also:

DRYING OILS. Adulteration indicated by lowered yield of INSOLUBLE BROMIDES.

COTTONSEED OIL. Halphen test, Titer test.

RAPE OIL. Lower saponification value,
Increased viscosity.

II. Technical.

Source.

The seeds of the FLAX PLANT (*Linum usitatissimum*).

Country of production.	Trade term of seed.	Percentage of production for 5 years ending 1912.
Argentina	River Plate	31·3
Russia	Baltic and Black Sea	23·7
United States	North America	20·3
India	East Indian	16·8
Canada[1]	Canadian	7·9

Content of oil in seed.

Variable with source of seed and season. Averages 35 to 38 per cent.[2].

Method of obtaining oil.

As the cake is of great value for feeding cattle the expression method is normally employed. The system used is the "Anglo-American" (see page 213).

COLD EXPRESSION is practised in Russia, Germany and India, and yields an edible oil of pleasant flavour and bright golden yellow colour.

HOT EXPRESSION. This is chiefly employed, as the main use of the oil is for commercial purposes. It yields a light brown oil, slightly turbid, owing to albuminous and extractive substances, and to moisture. The yield is 25 to 28 per cent.

Press cakes retain about 10 per cent. of oil.

Refinement.

"Green" or fresh seed yields more "spawn" than matured seed. The "spawn" or "mucilage" contains Phosphates of Lime and Magnesia. On settlement or "tankage" the oil gradually clears with subsidence of the impurities. "Tanked" oil is purer, and on this account more suitable for varnish making than oil chemically refined.

A more usual method, because more rapid, is the treatment of the oil with 1 to 2 per cent. of strong sulphuric acid. The oil darkens, and the charred matter carries down the substances in suspension, leaving a clear product. The best strength of acid depends on the amount of moisture and "spawn" present, but is usually about 88 to 92 per cent. H_2SO_4. A good agitation must be kept up, and a lead lined vessel with mechanical stirrer is preferable.

BLEACHING is said to be further obtained by the use of zinc, calcium and magnesium peroxides, about one pound being sufficient to bleach a ton of oil.

[1] Canada is rapidly increasing its production.
[2] East Indian oil sometimes contains up to 43 per cent.

ARTISTS' OILS are bleached by exposure to sunlight in shallow trays under glass, the oil having been usually obtained by cold expression. Oils intended for soft-soap making are improved by treatment with a small quantity of caustic alkali[1]. A slight bleaching and deodorising effect is produced by this means.

Commercial grades.

Linseed oil being mainly used for paints and varnishes, the relative drying qualities of the oils from different sources are important. The variation which occurs in commercial oils is mainly due to the admixture of other seeds with the linseed, such as mustard, hemp and rape, owing to these plants growing in the flax plantations.

The arrivals of Linseed into this country are tested by the "Incorporated Linseed Association" (foreign seeds, if oil bearing, being reckoned at half value of Linseed). The purest seed is the "Baltic." "Black Sea" contains up to 5 per cent. admixture. "River Plate" seed is fairly pure, but the oil therefrom does not equal the best Russian, probably owing to climatic differences. "East India" seed resembles "Black Sea" in containing foreign admixtures.

Properties and Uses.

Owing to its high content of glycerides of unsaturated acids, linseed oil, when spread in a thin film, rapidly absorbs oxygen from the air, becoming transformed into a solid elastic substance ("linoxyn"). This product is highly resistant to almost all chemical agents, and is insoluble in all oil solvents.

It is this property, and the brightness and elasticity of the skin so formed, that makes linseed oil so valuable as a vehicle for pigments in paints, and for varnishes. [The manufacture and properties of "Boiled Oil," and "Lithographic Varnish" will be described in a later section.]

Linseed oil is also used as an ingredient of "Putty" and in the manufacture of "Linoleum." It is one of the finest oils for soft-soap manufacture, and for this purpose oils of inferior drying properties are no disadvantage. The oil is now frequently deglycerinated, and its resultant fatty acids employed as soft soap stock. They contain from 83—94 per cent. of free fatty acids. Unlike that made from Cottonseed oil, linseed oil soft soap does not "fig" (i.e. the Potassium salts of the fatty acids do not crystallise out) in cold weather, which fact commends its use in many cases. The oil is extensively "hardened" by hydrogenation with nickel catalyst, and the stearine produced is of very fine quality.

[1] See Cottonseed oil.

§ 108. CHINESE WOOD OIL (TUNG OIL).

I. General and Analytical.

Character.

The COLOUR is pale yellow to dark brown.

Distinctive ODOUR, somewhat unpleasant.

STEARINE none at ordinary temperatures.

Physical and Chemical Data.

	Test	Average figure	Usual variations	Extremes recorded
(1)	Specific gravity at 15·5° C.	**0·941**	·941—·942	·933[1]—·9440[2]
(3)	Solidifying point fatty acids	**37°**		
(4)	Refractive index 40° C.	**1·475**		
(6)	Solubility. True Valenta	46·5°[3]		
	$\frac{V \times 10}{80}$	5·8°[3]		
	Alcohol reagent	75·8°[3]		
(8)	Iodine value	**165**	160—170	149[2]—176[2]
	(*a*) Bromine thermal test	**24·8**		
	(*b*) Maumené			
	(*c*) Livache			
(10)	Saponification value	**192**	190—193	
(11)	Insoluble bromide value	**0**		
(14)	Acid value	usually under 12		
(15)	Unsaponifiable matter (per cent.)		0·4—1·3	

Remarks.

Highest SPECIFIC GRAVITY of any fatty oil.

Highest known REFRACTIVE INDEX.

On drying the oil forms an OPAQUE and NON-ELASTIC skin.

Special Tests.

1. The oil on heating to about 300° to 310° C. for about 10 minutes is converted into a stiff jelly, non-fusible, and insoluble in fat solvents.

2. Iodine causes also solidification of Tung oil. A 20 p.c. solution of the oil in chloroform is mixed with an equal volume of a saturated chloroformic solution of Iodine. A stiff jelly is produced after 2 minutes.

Chemical Composition.

Tung oil contains a large proportion of the glyceride of ELEOMARGARIC ACID (see page 48), a stereo-isomeride[4] of Linolic Acid. Fahrion states that it contains 2 to 3 p.c. solid acids and 10 p.c. oleic acid. It yields no insoluble bromides and its drying properties are therefore due to the large amount of "tetra" unsaturated fatty acid (eleomargaric acid).

The solidification of Tung oil on heating is almost certainly due to **polymerisation**[5] of the glyceride of this acid.

[1] Nash. [2] Chapman. [3] Fryer and Weston.
[4] See chap. III, page 45 on stereoisomerism. [5] See page 23 on polymerisation.

Adulteration.

The value of Tung oil is mainly ruled by Linseed oil. When the latter is high in price, Tung oil may be adulterated with cheaper oils.

For the detection of Mineral oils, Rosin oils, Rosin and Marine oils, see under Linseed oil (page 115). One of the most likely adulterants would be SOYA BEAN OIL, as this is a product of the same country. Examination here would reveal lowered specific gravity, Refractive Index and Iodine Value. For the detection of small quantities of other oils, *Bacon's Test* is recommended.

Bacon's Test.

Into four test-tubes of equal size place equal volumes of:—

1. Pure Tung oil.
2. ,, ,, with 5 per cent. of vegetable oil (e.g. Soya Bean oil).
3. ,, ,, with 10 ,, ,, ,, ,,
4. Sample to be tested.

Immerse all four tubes in an oil bath at a temperature of 288° C. for 9 minutes. Remove tubes and stab jelly formed with a bright spatula.

Pure oil gives a clean hard cut.

With 5 per cent. of added oil, the cut is softer with feathered effect.

With 10 per cent. of added oil, the jelly is softer still, and over 12 per cent. will prevent gelatinisation altogether.

Stillingia oil is detected by its optical activity.

II. Technical.

Source.

The seeds of *Aleurites Cordata*, contained in a nut, and growing in China and contiguous countries. (A similar tree is native to Japan, and yields the Japanese Tung oil, slightly varying from the Chinese oil.)

Content of Oil.

About 53 per cent. in seeds.

Method of obtaining Oil.

Crude native methods are still in vogue. Wooden presses are used. The COLD PRESSED oil is light in colour and is mainly exported. The oil expressed hot has a very dark colour. The seeds are roasted over a naked flame and ground between stones prior to expression. Yield about 40 per cent. The oil cake is poisonous, and is used as a fertiliser.

Refinement.

No refinement is undertaken except the settlement and filtration of the oil by the merchants who collect it from native sources.

Commercial Grades.

The cold pressed "white tung oil" (mainly exported) and "black tung oil," which is hot pressed.

Properties and uses.

The oil is an important substitute for Linseed oil when the latter is high in price. At other times exports of Tung oil to Europe show a large falling off. In the year 1906, nearly 29,000 tons were exported from

Hankow. It is the quickest drying oil known, but is inferior to Linseed oil on account of the opacity and inelasticity of the skin produced.

Used in China as a natural wood-varnish and waterproofing material.

§ 109. Candlenut Oil.

An oil obtained from the seeds of *Aleurites moluccana*, a tree covering large areas in the western tropics. The fruits resemble nuts. The oil is obtainable in very large quantities and is suitable both for paints and varnishes, and as a soap-making oil. It has purging properties.

Data:

(1)	Specific gravity at 15·5° C.	·925[1]
(3)	Solidifying point of fatty acids	13° C.[2]
(4)	Refractive Index (Z. B.)	65·7[1]
(8)	Iodine value	164[1]
(10)	Saponification value	193[1]
(11)	Insoluble Bromide value	11·2—12·6[1]
(15)	Unsaponifiable (per cent.)	0·5—1·0[1]

§ 110. Hemp seed Oil.

The oil is obtained from **Hemp seed** (*Cannabis sativa*) cultivated in Western Europe, N. America, India and Japan. It has a greenish colour, and has been used extensively for the manufacture of green soft soaps. It is also suitable as a paint oil.

Data:

(1)	Specific gravity at 15·5° C.	·927[3]
(3)	Solidifying point of fatty acids	16°[2]
(8)	Iodine value	148[1]
(10)	Saponification value	191[1]

§ 111. Walnut Oil.

Obtained from the common walnut (*Juglans regia*) which contains over 63 per cent. of oil. The cold pressed oil has an agreeable taste and odour of walnuts. Hot pressed, it is greenish in colour with an acrid flavour. It is used as an artist's oil, the paint so made having less tendency to crack than in the case of linseed oil. Adulteration with Linseed oil is indicated by high Iodine value and a high yield of Insoluble Bromides.

Data:

(1)	Specific gravity at 15·5° C.	·926
(3)	Solidifying point of fatty acids	16°
(4)	Refractive Index (Z. B.)	65 to 67·5
(8)	Iodine value	140 to 150
(10)	Saponification value	193 to 196
(15)	Unsaponifiable matter	0·5 to 1 per cent.

[1] Lewkowitsch. [2] De Negri and Fabris. [3] Chateau.

§ 112. SOYA BEAN OIL.

I. General and Analytical.

Character.

COLOUR—brown; bleached oil yellow.

ODOUR—slight, somewhat distinctive.

STEARINE—none at ordinary temperatures.

Special Tests.

None.

Physical and Chemical Data.

	Test	Average figure	Usual variations	Extremes recorded
(1)	Specific gravity 15·5° C.	·925	·924—·926	
(3)	Solidifying point fatty acids	21·5°		
(4)	Refractive index 40° C. (Z. B.)	63		
(6)	Solubility. True Valenta	53·3°[1]		
	$\frac{V \times 10}{80}$	6·7[1]		
	Alcohol reagent	67·0°[1]		
(8)	Iodine value	130	126—135	
	(*a*) Bromine thermal test	23		
	(*b*) Maumené	90	88—92	
	(*c*) Livache			
(10)	Saponification value	192	190—192·5	
(11)	Insoluble bromide value	7·8 E. and M.		
(14)	Acid value	0·3	seldom over 5	
(15)	Unsaponifiable (per cent.)	0·5		

Chemical Composition.

According to Matthes and Dahle[2], it contains:

Palmitic Acid	15 per cent.	Linolic Acid	19 per cent.
Oleic Acid	56 „	Linolenic Acid about	4·8 „

Adulteration.

Not commonly practised. There is usually little inducement owing to the value being normally below that of other seed oils. Cotton seed is readily detected by the HALPHEN reaction, and Linseed by its increased Iodine, and Insoluble Bromide values.

II. Technical.

Source.

The Soya Bean (*Glycine hispida* and varieties) is native to China, Manchuria and Japan, and the plant is being cultivated in other countries. Although unknown in this country before the year 1908, enormous shipments of the beans have been made in recent years.

[1] Fryer and Weston. [2] *Archiv. d. Pharm.* 1911, 249, 424.

Content of Oil in Seed.

About 18 per cent.

Method of obtaining Oil.

Native methods are very crude, but the yield of oil is good, the oil being expressed in wooden presses for a prolonged period. Prior to expression, the bean is soaked and crushed in water, and then boiled with water. Yield about 13 per cent.

Modern machinery involves the use of Anglo-American presses[1], and the yield is about 10 per cent.

During periods of high prices for cotton seed and linseed oils, soya bean oil is extracted with volatile solvents, a greater yield being thus obtained.

The cake after expression is a very valuable cattle food, being indeed esteemed higher than cotton cake, on account of the apparently increased amount of cream in the milk of cows fed with it. The extracted meal is unsuitable for fodder.

Refinement.

The crude oil is frequently refined by means of caustic soda, as described under Cotton seed (page 126). This is especially the case when it has been obtained from "off" seed.

FULLER'S EARTH is used for bleaching the oil for edible purposes[2].

The oil is also bleached by means of Chlorine, using e.g. Dichromate of Soda and Hydrochloric Acid.

Properties and Uses.

The oil became popular as a soap-making material at a time when Cotton seed and Linseed were high in price. It is an ideal oil for soft-soap manufacture, having all the virtues of Linseed oil, and yielding a soap of a firmer texture.

The oil is now used in large quantities as an edible oil. It is also "boiled" for use in paints, usually together with linseed oil, and can be employed as a substitute for the latter in "linoleum."

On deglycerination, the soya bean fatty acids are usually green in colour, yielding however a yellowish brown soap on saponification with alkalies.

§ 113. POPPY SEED OIL.

I. General and Analytical.

Character.

Pale straw to golden yellow COLOUR.

ODOUR scarcely noticeable.

Good oil has a pleasant FLAVOUR.

STEARINE none.

[1] See page 213.
[2] Use of Fuller's Earth, see page 218.

Special Tests.

Bellier's test for Poppyseed oil in Walnut oil.

Chemical Composition.

Probably Solid Acids 7 per cent.[1] Linolic about 65 per cent.[2]
Linolenic under 5 per cent.[2] Oleic „ 30 „ [2]

Physical and Chemical Data.

	Test	Average figure	Usual variations	Extremes recorded
(1)	Specific gravity at 15·5° C.	**·925**	924—927	
(3)	Solidifying point fatty acids	**16°**	15·5°—16°	
(4)	Refractive index 40° C. (Z. B.)	**63·5**		
(6)	Solubility. True Valenta	53·0°[3]		
	$\frac{V \times 10}{80}$	6·8[3]		
	Alcohol reagent	67·0°[3]		
(8)	Iodine value	**134**	131—137	
	(*a*) Bromine thermal test	**22**	22—23	
	(*b*) Maumené	**87**	86—88	
	(*c*) Livache			
(10)	Saponification value	**192·5**	191—195	
(11)	Insoluble bromide value	nil	—	—
(14)	Acid value.	**1·5**	Not usually over 20	
(15)	Unsaponifiable (per cent.)	**0·5**	—	—

Adulteration.

Sesamé Oil is most frequent: shown by the Baudouin Test (see practical part).

Other drying oils by Insoluble Bromide Value (negative with Poppyseed) Cotton by Halphen reaction. See also under Linseed Oil (page 114).

II. Technical.

Source.

From the seeds of the Opium Poppy (*Papaver somniferum*) grown in India, Russia, France, Asia Minor, etc.

Content of oil in seeds.

45—50 per cent.

Method of obtaining oil.

The oil is twice expressed, the first treatment being "cold drawn." Yield about 38 per cent. This oil is pale golden yellow, and is the "white Poppyseed oil" of commerce.

The second (hot) expression yields about 3 per cent. of a reddish oil.

Refinement. Sun bleached for artists' use.

[1] Tolman and Munson, *Jour. American Chem. Soc.* 1903, 690.
[2] Hazura and Grüssner. [3] Fryer and Weston.

Properties and Use.

The best qualities are used as salad oils and for adulterating olive oil.

It is largely used for the preparation of artists' paints. Only the lowest qualities are available for soap-making.

§ 114. Nigerseed Oil.

This oil occasionally becomes available on the English market. It is obtained from Tropical Africa, and is golden yellow oil of pleasant odour and flavour. Iodine value 128—134. True Valenta 54°.

Used as an edible oil, and when cheap, as an excellent soft soap-making oil.

§ 115. Sunflower Oil.

From the seeds of the common Sunflower, largely cultivated in Russia, China, Hungary and India. The seeds contain nearly 50 per cent. of oil, which has a golden yellow colour and pleasant flavour. The cake is a valuable cattle food. The oil is largely used in the countries of origin for edible purposes. Hot pressed oil is acid refined, in a similar manner to Linseed oil (q.v.). Iodine value about 130. True Valenta 61·0°.

THE VEGETABLE SEMI-DRYING OILS

Typical oil, Cottonseed (p. 128)

CHAPTER VIII

CLASS III. VEGETABLE SEMI-DRYING OILS

Oils containing notable proportions of glycerides of acids which absorb 4 atoms of Bromine, and only small amounts (if any) of glycerides of acids absorbing 6 atoms of Bromine.

TYPICAL OIL. COTTONSEED.

§ **116.** The IODINE VALUES of these oils range approximately FROM ABOUT 120 TO 100. They form an intermediate class between the drying oils and the oils which have no drying properties at all. This is shown by the fact that they form a skin when exposed to the air at temperatures above the normal.

The important physical and chemical data are:

(**1**) Specific gravity.
(**3**) Solidifying point of fatty acids.
(**4**) Refractive Index.
(**5**) Viscosity.
(**6**) Solubility ("true Valenta")[1].
(**8**) Iodine Value.
(**10**) Saponification Value.
(**14**) Acid Value.
(**15**) Unsaponifiable matter.

The most important determination is the IODINE VALUE (**8**).

The OPTICAL ACTIVITY (7) is virtually negative as in the case of the drying oils. The ELAIDIN TEST (9) yields a product of BUTTERY consistency. The INSOLUBLE BROMIDE VALUE (11) is negative in all cases, and as the proportion of glycerides of fatty acids of low molecular weight is very small in fresh oils, the Reichert-Meissl test (12) is useless for discriminative purposes.

§ 117. MAIZE OIL.

I. General and Analytical.

Character.

COLOUR—golden yellow.

FLAVOUR AND ODOUR—distinctive; retains that of the Indian Corn.

"STEARINE"—slight yield at ordinary temperatures.

[1] See footnote, page 75.

Physical and Chemical Data.

	Test	Average figure	Usual variation	Extremes recorded
(1)	Specific gravity at 15·5° C.	**·925**	921—927	
(3)	Solidifying point of fatty acids	**18·5°**	18°—19°	
(4)	Refractive index 40° C. (Z. B.)	**60·3**	59·5—60·5	
(6)	Solubility. True Valenta	68·0°[1]		
	$\frac{V \times 10}{80}$	8·5[1]		
	Alcohol reagent	68·0°[1]		
(8)	Iodine value	**119·2**	115—125	
	(*a*) Bromine thermal test	**21·5**		
	(*b*) Maumené	**82**	79—86	
	(*c*) Livache			
(9)	Elaidin	buttery		
(10)	Saponification value	**191·5**	190—193	
(14)	Acid value	**6**	"old process oil" high [up to 30 per cent.]	
(15)	Unsaponifiable matter	**1·5**		

Remarks.

The old figures obtained for Reichert-Meissl and Acetyl Values were due to decomposition of oil during fermentation of the seed. The Unsaponifiable matter is high on account of the **lecithin**[2] contained in the oil.

Chemical Composition.

Solid acids, probably Palmitic and Arachidic apparently about 7·5[3] per cent.; Liquid acids, a mixture of Linolic and Oleic.

Adulteration.

Cottonseed oil only likely adulterant (when cheaper in price). Detected by HALPHEN TEST (see later section) and increased TITER (3).

II. Technical.

Source.

The "germs" of the Indian Corn (*Zea Mais*).

Content of Oil.

Dry germs contain about 53 per cent. oil.

Method of obtaining Oil.

The oil is a bye product in the production of starch. Formerly it was produced in alcohol distilleries. Owing to the fermentation, a very dark acid oil was obtained.

The present method consists in

1. Malting.
2. Crushing.
3. Freeing the "Germ" from starch by washing with water and straining.
4. Germs expressed in oil presses.

Yield from dry germ 40 per cent.

The cake is utilised as fodder for cattle.

[1] Fryer and Weston. [2] See page 14. [3] Tolman and Munson.

Refinement.

Unnecessary in most cases. Dark qualities are ALKALI refined. (See Cottonseed oil, page 130.)

Properties and Uses.

On account of its too-distinctive flavour, it is mixed with other oils for use as a salad oil and in the preparation of margarine. It is an excellent oil for soft soap-making.

§ 118. Kapok Oil.

From seeds of trees related to the cotton plant which are very abundant in tropical regions. The oil gives the characteristic **Halphen** test for Cottonseed oil, but is said not to respond to the **Becchi** test. Iodine value 111—120. Other data very close to those of Cottonseed. Sold in place of latter and in admixture with it.

§ 119. TYPICAL OIL. COTTONSEED OIL.

I. General and Analytical.

Character.

COLOUR. Crude—ruby red. Refined—golden yellow.

ODOUR AND FLAVOUR. Somewhat pronounced and distinctive. Odour very distinctive on acidifying saponified oil.

STEARINE. Deposits fair amount. "Winter" oils are destearinated.

Physical and Chemical Data.

	Test	Average figure	Usual variations	Extremes recorded
(1)	Specific gravity at 15·5° C.	·923	·922—·926	
(3)	Solidifying point fatty acids	35°	33°—37° natural oil	
	" " "	28°	28°—28·5° winter oil	
(4)	Refractive index 40° C. (Z. B.)	58·5	58—59	
(6)	Solubility. True Valenta	59·5°[1]		
	$\frac{V \times 10}{80}$	7·4[1]		
	Alcohol reagent	65·2°[1]		
(8)	Iodine value	110	105—112	101[2]—120[3]
	" liquid fatty acids	—	147—151	141[4]—
	(*a*) Bromine thermal test	19·4		
	(*b*) Maumené	78		75[5]—90[2]
	(*c*) Livache 24 hrs. exposure	5·9		
(9)	Elaidin	buttery		
(10)	Saponification value	193	192—194	
(14)	Acid value	usually not over 0·5		
(15)	Unsaponifiable	·9	0·7—1·7	

[1] Fryer and Weston. [2] Wiley. [3] Lewkowitsch. [4] Bömer. [5] Archbutt.

The most important tests for the recognition of Cottonseed oil are the TITER TEST (**3**) and IODINE VALUE of liquid fatty acids.

Special Tests.

1. Halphen test—crimson coloration.
2. Nitric acid test—coffee-brown coloration.
3. Becchi test—darkening occurs.

NOTE. The Halphen test is given by Kapok oil, and is rendered negative by heating the oil to high temperatures. It is also not given by a hydrogenated oil. Animals fed on cottonseed cake yield an oil or fat giving this reaction.

Chemical Composition (see remarks on Lard, page 164).

Solid acids 20—25 per cent. consisting mainly of Palmitic acid (little Arachidic ?)[1].

Liquid acids, proportion not yet agreed. Probably about 30 per cent. Linolic and remainder Oleic.

Adulteration uncommon. Linseed oil is detected by higher Iodine value and Insoluble Bromide value. Soya oil also gives small yield Hexabromides. Maize oil adulteration is difficult to detect. Experts can tell by the flavour.

II. Technical.

Source.

The seeds of the cotton plant; several species of Gossypium.

Countries of Origin.

United States, Egypt, Sea Islands, India (Bombay) and Levant. Some seed is also produced in South America, Russia and West Africa.

Content of oil in seeds.

About 20 per cent.

Method of obtaining oil.

Expression in hydraulic presses is the universal rule owing to the value of cotton cake for cattle fodder.

The cotton seed is covered with the well-known woolly hairs constituting the raw cotton[2]. In the case of Egyptian and Sea Island seed, this is fairly easily removed from the seed, but the American, Levant and Bombay varieties retain the fibres so tenaceously that the seeds cannot be entirely freed from them.

On this account, American, Levant, and East Indian seed is "decorticated"—that is, the husk or "hull" is removed—previous to expression. The machine for this purpose is termed the "huller" (see illustration Plate XII). The hulls and meats are then separated one from the other (see Plate VIII) and the meats pressed for oil.

[1] See V. J. Meyer, *Chem. Zeit.* 1907, 794.

[2] See Plate VII.

The Egyptian and Sea Island seed is expressed in the same manner as Linseed, after removing adhering hairs in a "delinter" (see Plate X). The seeds are made into meal in 'crushing rolls,' cooked, and expressed once in a press of the Anglo-American type (see page 213).

The press cakes contain about 10 per cent. of oil, which is intentionally retained owing to its value in the cattle cake.

As the kernels contain a reddish brown colouring matter, the crude oil, after expression, is of a deep ruby red colour.

Refinement.

The crude oil is usually refined in the mills. This is accomplished in a very simple manner, but one which requires great care to prevent loss. The crude oil is warmed, and then agitated with a dilute solution of caustic soda in sufficient quantity only to combine with the colouring matter and with the free fatty acids. Samples are taken, and when they show (on a piece of glass) a distinct "break," i.e. separation between the oil and soda liquor, the agitation is stopped.

The soap formed by the interaction of the free acid and the caustic carries down with it practically all the colouring matters, forming a thick residue ("foots" or "mucilage") and leaving the supernatant oil of a yellow to brownish yellow colour. The "foots" which contains oil entangled by the soap, is either made into soap stock, by saponifying and "cutting" the soap with excess of alkali, or is acid treated, and the separated black oil distilled yielding fatty acids of light colour which are pressed for "stearine" and "oleine."

The cotton oil is then washed, and, if intended for edible purposes, is bleached with Fuller's Earth[1], a chemical bleach being unsuitable. Oil for industrial purposes is bleached with bleaching powder and sulphuric acid (see p. 223).

Commercial Grades.

The crude oil and the refined oil are graded in America as "Choice," "Prime" and "Off," the latter being sold on sample. In this country, oil is quoted on the market as "Crude," "Refined" and "Ordinary Pale." The latter is the soap-making grade.

"Winter Oils" are prepared by separating the solid fat, which is then known as "cottonseed stearine."

Properties and Uses.

Cottonseed oil, originally employed mainly as a soap-making oil, has been increasingly utilised for a raw material for edible prepara-

[1] See page 218 and Plate XXXVII.

tions, such as margarine, cooking fats, salad oils, etc. On this account, its price has in recent years been much against its use as a soft-soap making oil, and the advent in the year 1908 of Soya Bean oil, contributed towards this result. As a hard-soap making oil the use of the lower grades is still probably as great as ever. When "blown," it is employed as a lubricant.

§ 120. SESAMÉ OIL.

I. General and Analytical.

Character.

COLOUR, light yellow. ODOUR, very faint.

FLAVOUR, best oils pleasant nutty.

"STEARINE," very small at ordinary temperatures.

Physical and Chemical Data.

		Average figure	Usual variations	Extremes recorded
(1)	Specific gravity at 15·5° C.	**·923**	·921—·925	
(3)	Solidifying point of fatty acids	**23°**	22°—24°	
(4)	Refractive index 40° C. (Z. B.)	**60**	59·5—60	40·6[1]— African oil
(6)	Solubility. True Valenta	70·0°[2]		
	$\frac{V \times 10}{80}$	8·7[2]		
	Alcohol reagent	68·5°[2]		
(7)	Polarimeter (200 mm.)	**+1·2°**	+·8° to +1·6°	
(8)	Iodine value	**105**	103—110	—115[3]
	(*a*) Bromine thermal test	**23·5**		
	(*b*) Maumené	**64**	63—65	
(9)	Elaidin	reddish brown semi-liquid		
(10)	Saponification value	**192**	188—193	
(14)	Acid value	**5**	varies with grade	
(15)	Unsaponifiable matter	**1·0**	0·8—1·2	

Special Tests.

Baudouin test gives crimson colouration of test liquid. Hardened ("hydrogenated") fats give reaction only faintly[4].

Chemical Composition.

Lane[5] finds	Solid Acids	about 12 per cent.	
	Linolic Acid	,, 16	,,
	Oleic ,,	,, 72	,,

The unsaponifiable matter consists of Phytosterol, SESAMIN a resinous body melting at 118°, and a brown oily body giving the characteristic Baudouin colour reaction.

[1] Wagner. [2] Fryer and Weston. [3] Shukoff.
[4] Paal and Roth, *Berichte* 1909, 1550. [5] *Jour. Soc. Chem. Ind.* 1901, 1083.

Adulteration.

Poppyseed, cottonseed, arachis and rape oil are likely adulterants. Best tests are: Poppyseed, Iodine Value; Cottonseed, Titer; Arachis, Arachidic acid test; Rape, Saponification Value.

II. Technical.

Source.

Varieties of the genus Sesamum yield a fruit containing numerous small black or white seeds, the latter yielding the finest oil. The best oil is obtained from the LEVANT variety; but the greatest quantity ot seeds is produced in India, China and Japan.

Content of oil in seed.

50—57 per cent.

Method of obtaining oil.

If the seed is in good condition, it is expressed. Unsound seed, and rejected cake is extracted with solvents, usually Petroleum Ether or Carbon Disulphide (see page 215). The seeds containing a large percentage of oil, are pressed two or three times. The first expression, in the cold, yields the best oil. Second and third expressions of the cooked seed yield lower qualities. The press for the first expression is of the cage type (see page 213). Subsequent expressions are usually made in the Anglo-American type.

The cake contains about 9 per cent. of oil, and is largely used for cattle feeding. Extracted meal is only suitable for manure.

Refinement.

The cold expressed oil after filtration is suitable for use. Lower qualities are bleached in the usual way for soap-making chiefly at Marseilles.

Properties and Uses.

Sesamé oil is chiefly employed as a constituent of edible fats and as a salad oil. In Germany and Austria the use of 10 per cent. and in Belgium of 5 per cent. of Sesamé oil (computed from the total fatty matters present) is obligatory in margarine in order to afford a ready means of recognition by the Baudouin Test. This accounts for a large quantity of oil. Lower qualities are used for soap-making, especially in France.

§ 121. Croton Oil.

This is an oil obtained from China and Southern Asia possessing very powerful purgative properties. It has a strong, disagreeable odour, and burning taste, and is used exclusively for pharmaceutical purposes. It is stated to contain Valeric and Butyric acids as glycerides, and gives a Reichert-Meissl value of 12—13. The purgative property disappears on "hydrogenating[1]." The Iodine value usually varies from 102—104.

§ 122. Curcas Oil.

Obtained from the "Purging nut" of South America, and like Croton oil possesses powerful purgative properties. It has a characteristic and unpleasant odour, and is used as a soap-making oil. Iodine value 98—100.

[1] *Berichte* 1909, 1546.

THE VEGETABLE NON-DRYING OILS

1. **Rape Oil Group**
2. **Olive Oil Group**
3. **Castor Oil Group**

CHAPTER IX

CLASS IV. VEGETABLE NON-DRYING OILS

Oils characterised by the preponderance of glycerides of acids absorbing 2 atoms of bromine to the practical exclusion of less saturated fatty acids.

§ **123.** At elevated temperatures these oils thicken and become more viscous in consistence, but they do not dry to a skin in the air even after prolonged exposure. On this account they are all highly suitable for use as lubricants, as they have little tendency to 'gum.'

The IODINE VALUES range from about 100 TO 80. They form themselves into three well-defined groups (of which however the last consists only of one member of importance).

Group 1. Rape oil group. Characteristic acid Erucic.
Group 2. Olive oil group. " " Oleic.
Group 3. Castor oil group. " " Ricinoleic.

The important chemical and physical data are:

(1) Specific gravity
(3) Solidifying point fatty acids
(4) Refractive index
(5) Viscosity (especially groups 1 and 3)
(6) Solubility[1] (especially groups 1 and 3)
(7) Polarimeter (group 3 only)
(8) Iodine value
(9) Elaidin
(10) Saponification value (especially group 1)
(13) Acetyl value (group 3 only)
(14) Acid value
(15) Unsaponifiable matter.

§ 124. Group 1. Rape Oil Group.

TYPICAL OIL—RAPE OIL.

As previously stated, the oils of this group are all characterised by the presence of the glyceride of **erucic acid** (page 47). As this acid contains **22 atoms** of carbon, as distinct from almost all the other oils (which contain acids having 18 or less carbon atoms[2]) the mean molecular weight of the mixed fatty acids is higher than normal, and

[1] See footnote, page 75. [2] See however Arachis oil, and some rarer oils.

consequently the **saponification value** is **lower than normal**. The determination of this figure is therefore the most important test for the oils of this group.

These oils are all obtained from plants belonging to the natural order Cruciferae (cabbage family).

§ 125. Ravison Oil.

This oil is from a wild variety of rape seed, also called "Black Sea rape." It is extracted from the seeds by volatile solvents, and owing to its higher iodine value (101–120) is not so suitable as a lubricant as true rape oil.

§ 126. TYPICAL OIL. **RAPE OIL** (COLZA OIL).

I. General and Analytical.

Character.

COLOUR—Light yellow and dark brown.

ODOUR—Well known and distinctive.

FLAVOUR—Rough and harsh, except finest qualities.

STEARINE—An appreciable quantity separates at ordinary temperatures.

CONSISTENCY—Viscous.

Physical and Chemical Data.

	Test	Average figure	Usual variations	Extremes recorded
(**1**)	Specific gravity at 15·5° C.	**·9155**	·9135—·917	
(**3**)	Solidifying point fatty acids	**12·5**	12—13	
(**4**)	Refractive index 40° C. (Z. B.)	**59**	59—60	
(**5**)	Viscosity $\eta \times 100$ (15·5° C.)	**112**		
(**6**)	Solubility. True Valenta	110·3°[1]		
	$\frac{V \times 10}{80}$	13·8[1]		
	Alcohol reagent	83·5°[1]		
(**8**)	Iodine value	**100**	97—105	
	(*a*) Bromine thermal test	**18·5**	17—20	
	(*b*) Maumené		55—64	
	(*c*) Livache			
(**10**)	Saponification value	**175**	170—177	167·7[2]—178·7[3]
(**11**)	Insoluble bromide value	**2—3**		
(**14**)	Acid value	not usually exceeding 4		
(**15**)	Unsaponifiable	**·75**	0·5—1·0	

Remarks. The VISCOSITY figure (highest known except Castor) as well as the SAPONIFICATION VALUE is important as a test of purity.

The VALENTA is an excellent sorting test.

The INSOLUBLE BROMIDE TEST distinguishes it from the other non-drying and most of the semi-drying oils.

[1] Fryer and Weston. [2] Crossley and Le Sueur. [3] Köttstorfer.

Special tests.

TORTELLI AND FORTINI's method for detection of rape oil in mixtures depends upon the presence of the solid unsaturated fatty acid (Erucic) in rape oil.

Chemical Composition.

Archbutt obtained from ·5 to 1·6 per cent. of a saturated acid (probably Arachidic). Rapic acid occurs as well as the characteristic ERUCIC ACID, and probably a small percentage of LINOLENIC or similar acid, since an insoluble bromide value of about 3 per cent. has been obtained by *Hehner and Mitchell.*

Dierucin often occurs in crude Rape oil[1], though not in refined oils. "Brassicasterol" is stated to occur in the unsaponifiable matter.

Adulteration.

Likely adulterants may be (according to market values of oils):

LINSEED—Iodine value, viscosity, insoluble bromide value.

COTTONSEED—Halphen test, solidifying point fatty acids.

MINERAL OILS AND ROSIN OILS—Unsaponifiable matter.

The viscosity is a good test of purity.

Adulteration with the allied varieties, Ravison and Jamba oils, may be practised and is next to impossible to detect by chemical means.

II. Technical.

Source.

From the seeds of *Brassica Campestris*, of which there are several varieties.

The plant is cultivated largely in Europe, and imported seed is obtained from India in huge quantities. Oil is also exported from Japan (usually in Petroleum oil cases).

Content of Oil in Seed.

40—45 per cent.

Method of obtaining oil.

By expression or extraction with volatile solvents. Expression in the cold ("cold drawn" oil) is employed for the finest qualities, but the seed is mainly pressed once hot. The cake is a good cattle food.

Extracted meal is suitable only for manure, but the best qualities are occasionally mixed with other meals in small quantities.

Refinement.

Rape oil is refined by sulphuric acid in a similar way to Linseed oil (see page 116). As free mineral acid is a very objectionable

[1] First discovered by Reimer and Will

constituent, especially for lubricating purposes, the oil should be examined for its presence (by shaking with hot distilled water).

Properties and Uses.

The "cold drawn" oils are used as edibles, especially in India. Less refined oils are still largely used for burning. On account of its high viscosity, and non-gumming properties, rape oil is used in enormous quantities as a lubricant, usually dissolved in mineral oils. To increase the viscosity the oil is "blown." It is not very suitable for soap-making, the soap produced being of poor "body."

§ 127. Mustard Oil.

Black and white mustard oils are obtained from *Sinapis nigra* and *S. alba* respectively. They closely resemble rape oil in composition, but are distinguished therefrom by the pungent odour of **ethereal mustard oil.**

Saponification value, about 173; iodine value: black, 105–110, white, 92–103. The black mustard oil is stated to be used chiefly for soap-making; the white for burning and lubricating.

§ 128. Jamba Oil.

Closely related to **rape oil,** but inferior thereto as it is difficult to obtain "blown" oil of satisfactory quality from it. Recognised, and distinguished from genuine rape oil by its distinctive flavour and odour.

§ 129. Group 2. Olive Oil Group.

TYPICAL OIL—OLIVE.

These oils are characterised by the presence of large proportions of the glyceride of **oleic acid**. They contain only small amounts of linolic and no linolenic acid. Owing to the preponderance of oleic acid, they yield **solid elaidins.**

§ 130. The Kernel Oils.

Cherry, Apricot, Plum and Peach Kernel oils, often cold expressed, are used mainly as adulterants for Almond oil. They may all be detected by *Bieber's Reagent* which gives colour reactions with them.

Iodine Values. Cherry K., 110—113; Apricot K., 101—108; Plum K., 91—100; Peach K., 92—109.

§ 131. ALMOND OIL.

I. General and Analytical.

Character.

COLOUR—almost water white.

ODOUR—slight.

FLAVOUR—almost tasteless.

"STEARINE"—nil.

Physical and Chemical Data.

	Test	Average figure	Usual variations	Extremes recorded
(1)	Specific gravity at 15·5° C.	**·918**	·9178—·9183	
(3)	Solidifying point fatty acids °C.	**10·5°**	9·5°—12°	
(4)	Refractive index 40° C. (Z. B.)	**57**	56·5—57·5	
(6)	Solubility. True Valenta	80·0°[1]		
	$\frac{V \times 10}{80}$	10·0[1]		
	Alcohol reagent	70·0°[1]		
(8)	Iodine value	**99**	98—100	93[2]—101·2[3]
	(*a*) Bromine thermal test	**20**	18—21	
	(*b*) Maumené	**52**	51—53	
(9)	Elaidin	solid		
(10)	Saponification value	**192**	189—193	
(14)	Acid value		0·5—5	
(15)	Unsaponifiable	**0·75**	0·5—1·0	

Special Tests.

The mixed fatty acids of almond oil, mixed with an equal volume of absolute alcohol at 15° C. should dissolve to a clear solution, and no turbidity should be observable on adding a further volume of alcohol. This test will detect the presence of olive, arachis, sesamé and cottonseed oils.

Chemical Composition. Consists almost entirely of the glyceride of oleic acid; but some linolic is also present. Stearic acid is absent.

Adulteration.

Special test as above. Important test is the solidifying point of mixed fatty acids which is very low. Chief adulterants are the kernel oils detected by Bieber's test (though not very satisfactorily).

II. Technical.

Almond oil is chiefly obtained by expression of bitter almonds. The press cakes are used in the preparation of the **essential oil**. The yield of oil is from 38 to 45 per cent. The oil is normally too dear for any other use than in pharmacy.

§ 132. ARACHIS OIL.

I. General and Analytical.

Character.

COLOUR—pale yellow to light brown.

ODOUR and flavour—"nutty."

"STEARINE"—small amount usually separates.

Special Tests.

Bellier's Test (qualitative only).

Renard's Test (isolation of arachidic acid).

[1] Fryer and Weston. [2] De Negri. [3] Allen.

Physical and Chemical Data.

	Test	Average figure	Usual variations	Extremes recorded
(1)	Specific gravity at 15·5° C.	**·9175**	·916—·918	
(3)	Solidifying point fatty acids ° C.	**28·5°**	28°—29°	
(4)	Refractive index 40° C. (Z. B.)	**55·5**	55—57·5	
(6)	Solubility. True Valenta	88·5°[1]		
	$\frac{V \times 10}{80}$	11·0[1]		
	Alcohol reagent	74·2°[1]		
(8)	Iodine value	**90**	88—95	83·3[2]—105[3]
	(*a*) Bromine thermal test	**16·5**		
	(*b*) Maumené		46—51	
(9)	Elaidin			
(10)	Saponification value	**193**	191—196	
(14)	Acid value	**10**, not usually over 20		
(15)	Unsaponifiable matter	**0·5—1·0**		

Chemical Composition.

It is uncertain whether Palmitic acid occurs in Arachis oil. The *characteristic* acid is ARACHIDIC which occurs to the extent of about 5 PER CENT. LIGNOCERIC acid also occurs. Oleic acid is the chief liquid acid, together with Linolic and probably HYPOGÆIC ACID[4].

Adulteration.

Sesamé—Baudouin test[5] (confirm with Iodine Value, Titer, Renard).
Cottonseed—Halphen test (Titer, Iodine Value, and of liquid fatty acids).
Poppyseed—Iodine Value, Specific gravity.
Rape—Saponification Value, Titer.

II. Technical.

Source.

The "Ground-nut" or "Pea-nut" (*Arachis hypogæa*), known familiarly as the "monkey nut." Widely cultivated in the Far East, West Africa, India, and North and South America.

Content of Oil in Seed. 43—46 p.c. Some varieties much less.

Method of obtaining Oil.

The shells are cracked and removed by special machinery. The skin is then removed from the kernel as completely as possible, and the kernels ground and pressed in a cage-type press in the cold. The yield of cold drawn oil is about 18 per cent. Usually two subsequent expressions are given at increased temperatures.

Refinement. The edible varieties are bleached to a very light colour by means of Fuller's earth or decolourising carbon.

For soap-making purposes it is alkali refined.

[1] Fryer and Weston. [2] Tortelli and Ruggeri. [3] Oliveri.
[4] Meyer and Beer, *Monatsch. f. Chem.* 1913, 1195.
[5] May contain slight Sesamé oil contamination from mill (Lewkowitsch).

Commercial Grades and Uses of Oil.

FIRSTS. Cold drawn oil. Used as a salad oil and for adulterating olive oil, also for packing tinned fish.

SECONDS. *Warm expression.* Refined for a lower grade edible and sometimes used for soap.

THIRDS. *Hot expression.* Chiefly soap-making (also burning).

§ 133. Rice Oil.

Extracted or expressed from **rice meal** (9—15 per cent. oil) or rice bran. A fair quantity is produced at Hull. The oil has usually a dark brown colour, and contains much "stearine" which is separated fatty acid, the percentage of free acid being very high. It has a peculiar musty odour, extremely difficult to remove. Iodine value 97, saponification value, 193.

§ 134. TYPICAL OIL. OLIVE OIL.

I. General and Analytical.

Character.

COLOUR—very light or golden yellow to dark dirty green.

ODOUR—varies from slight to strong and somewhat pungent and is characteristic with freshly-liberated fatty acids.

FLAVOUR—mild to acrid (of edibles finest flavour is *Tunisian*).

STEARINE—variable. Salad oils free.

Physical and Chemical Data.

	Test	Average figure	Usual variations	Extremes recorded
(1)	Specific gravity 15·5° C.	**·916**	·915—·918	
(3)	Solidifying point fatty acids °C.	**20°**	18°—22°	16·9°[1]—26·4°[1]
(4)	Refractive index 40° C. (Z. B.)	**55·5**	54·5—56·5	
(5)	Viscosity 15·5° C. ($\eta \times 100$)	**100·8**		
(6)	Solubility. True Valenta	76·5°[2]		
	$\frac{V \times 10}{80}$	9·5[2]		
	Alcohol reagent	69·0°[2]		
(8)	Iodine value	**83**	80—85	77[3]—93·7[4]
	(*a*) Bromine thermal test	**14·5**	13·5—15	
	(*b*) Maumené	**46**	45—47	
(9)	Elaidin	hardest of all oils, in shortest time		
(10)	Saponification value	**190·5**	190—195	185[5]—
(14)	Acid value	**0·5** to **50**	(edibles 0·5 to 2)	
(15)	Unsaponifiable matter p.c.	**0·75**	0·5 to 2	

Special Tests. None[6].

[1] Lewkowitsch. [2] Fryer and Weston. [3] Lengfeld.
[4] Crossley and Le Sueur. [5] De Negri and Fabris.

[6] A new test has been proposed by *Mazzaron* (see p. 82) depending on the estimation of the sulphur dioxide evolved on mixing with Sulphuric Acid (as in Maumené test). The figures given vary from 2·4 (Olive) to 223 (Soya).

Chemical Composition.

Solid acids 15 to 25 per cent.

Consisting of PALMITIC and a very small quantity of Arachidic acids.

The liquid acids are preponderantly Oleic with approximately 7 per cent. of Linolic acid.

Rancid oils contain aldehydes and volatile acids, such as acetic, butyric, œnanthylic, etc.

Adulteration.

Apart from the admixture of other oils, which is largely practised, there is the question of the misbranding of oils, and admixture of oils of lower quality with fine edible oils. Recently much "sulphur extracted" olive oil has been passed off for a higher quality. These oils can apparently be recognised by experts by the appearance of a slight bloom in reflected light over a dark background. (See also Halphen's and Milliau's tests for carbon disulphide[1], in practical part.) The acid value was formerly considered a test for the quality of an oil, but as free acid is now removed by means of alkali, this no longer applies. Such oils however would show a *lower* acid value than normal.

Adulteration with foreign oils is to be ascertained only by a careful determination of the constants given above. The IODINE VALUE is particularly important, inasmuch as almost any oil used as an adulterant would be higher than Olive.

The Iodine Value appears to vary in genuine olive oils according to the degree of ripeness of the fruit, the ripest fruit giving the highest values[2].

The colour tests for Cotton and Sesamé oils are important, in case high Iodine values are obtained. ARACHIS oil is recognised by Bellier's test and if present the percentage determined by Renard's test.

Lard oil yields characteristic crystals, and is confirmed by the melting point of the STEROL ACETATE.

II. Technical.

Source.

From the well-known fruits of the Olive tree, *Olea europea*, grown in Spain, Italy, South France and Northern Africa.

Content of oil in fruit.

Variable; 35 to 60 per cent.

[1] As little as **0·02 per cent.** of carbon disulphide can be detected by the Halphen reaction using *Kapok* oil.

[2] Paparelli, *Jour. Soc. Chem. Ind.* 1892, 848.

Method of obtaining oil.

In the smaller mills the kernels are not removed. Larger works remove these before crushing the fruits.

A. The hand-picked fruits are crushed in a mill and the pulp expressed in a hydraulic press (Marseilles or "Pack" press, see page 213). This gives edible oil of the finest grade.

B. The pulp (or "marc") is mixed with cold water and repressed, yielding an edible oil of a like superior quality to (A).

C. It is now expressed again after admixture with hot water. This yields a commercial oil, which is however often passed, after refinement, for edible purposes.

D. Two methods are now employed. The pulp (i) finely ground in special mills with water, and run into tanks, where a further yield of oil rises to surface; (ii) allowed to ferment in pits in a similar manner to "Palm Oil" (q.v.). In either case the oil contains a high proportion of free fatty acids.

E. The pulp is finally extracted, after drying, by means of carbon disulphide yielding the "Sulphur Olive oils." These are of a dark green colour, and possess a disagreeable "oniony" smell.

Refinement.

Cold expression oils are brightened by filtration but are not usually further treated, except to remove STEARINE.

Lower quality oils are acid refined with sulphuric acid, and in some cases the free fatty acids are removed by means of caustic alkalis.

Properties and Uses.

The finest oils are the SALAD OILS *par excellence* These should be almost free from fatty acids though a small percentage is said to improve the flavour. Oils vary very widely due to:

(1) Variety of tree (Italy produces 300 varieties),
(2) Degree of ripeness of fruit,
(3) Method of gathering,
(4) Method of expressing.

Attention is paid to the condition of the fruit, the skin of which must be unbroken. The oil must also be expressed as soon after gathering as possible. The Tunis olives yield the best flavoured varieties.

Lower qualities are used chiefly for soap-making. The official B.P. soft soap is made from low grades of Olive oil; as are also the finest textile soaps. It is also employed to some extent for burning and lubricating.

§ 135. Group 3. Castor Oil Group.

TYPICAL OIL—CASTOR.

Characterised by the presence of a glyceride of a HYDROXYLATED fatty acid, e.g. *ricinoleic acid.*

CASTOR OIL.

I. General and Analytical.

Character.

COLOUR—water white to dirty green.

ODOUR—slight.

FLAVOUR—characteristic, "rough."

STEARINE—none.

CONSISTENCY—very viscous.

Physical and Chemical Data.

	Test	Average figure	Usual variations	Extremes recorded
(1)	Specific gravity 15·5° C.	**·964**	·960—·967	
(3)	Solidifying point fatty acids	**3°**		
(4)	Refractive index 40° C. (Z. B.)	**68·9**	60—70	
(5)	Viscosity 100° F. ($\eta \times 100$)	**272·9**	—	—
(6)	Solubility	soluble acetic acid, alcohol. Insoluble petrol ether		
(7)	Polarimeter 200 mm.	**+8·8**	+7·5 to +9	
(8)	Iodine value	**88**	82—90	
	(*a*) Bromine thermal test	**14·7**	14·5—15	
	(*b*) Maumené	**46·5**	46—47	
(9)	Elaidin	pearly solid		
(10)	Saponification value	**180**	176—184	
(13)	Acetyl value	**150**	147—150	
(14)	Acid value	not over **4** °/₀		—30·5[1]
(15)	Unsaponifiable	**0·3** to **0·6**		

Remarks. In the elaidin test, the Ricinoleic is changed to Ricinelaidic acid. The oil is characterised by its *high specific gravity, viscosity* and *optical rotation.* Also by the *acetyl value* and *solubility in alcohol.*

The optical activity is caused by the asymmetric carbon atom of Ricinoleic acid (see page 52).

Chemical Composition.

Chiefly the glyceride of *Ricinoleic acid*; a small quantity of hydroxystearic acid and of stearic acid also occurs.

[1] Atlas ("thirds"—in author's laboratory).

Adulteration.

Other vegetable oils. All lower the acetyl value, rotation, and viscosity, and are thus readily discovered.

II. Technical.

Source.

From the seeds of *Ricinus communis*, grown in India, and in most hot countries.

Content of Oil in Seed.

46—52 per cent.

Method of obtaining Oil.

By expression and extraction.

The latter method is unsuitable for medicinal oil. The seed is first decorticated (see plates VII, XI). The first expression is in the cold, and yields about 33 per cent. of medicinal oil. (Cage form of press used, see page 213.) Two further hot expressions are then made and the cake is then extracted by carbon disulphide after grinding. The exhausted meal, being poisonous, is useless for cattle feeding, and is employed as manure.

The oil is extracted from the whole seed in Hull for manufacturing purposes.

Refinement.

Chiefly by steaming, which coagulates albuminous matters, followed by filtration.

Properties and Uses.

Being a mild purgative, it is very largely employed for medicinal purposes. Only the cold drawn oil is suitable.

Large quantities are used for lubricating in blended lubricating oils, and especially in marine oils. It is employed to give transparency to hard soaps. A large amount is also made into "soluble" or "turkey-red" oil. There are many other minor uses to which the oil is put in commerce.

The *Seed* contains a fat splitting enzyme which is extracted and used commercially for this purpose, although of late years the price of the seeds has militated much against its extended use.

THE ANIMAL OILS

Typical oil, Neatsfoot

CHAPTER X

CLASS V. ANIMAL OILS

§ 136. Typical Oil—Neatsfoot.

These oils—to be distinguished from the fish oils and the oils from the fish-eating mammals—are *chemically* a sub-class of the non-drying vegetable oils. Thus, they consist very largely of the glyceride of acids which absorb only **2 atoms of bromine** (i.e. **oleic acid**) and contain little or no acid more unsaturated than oleic. As however they all contain **cholesterol** in the unsaponifiable matters, and can thus be readily distinguished from the non-drying vegetable oils, it is convenient to group them as a distinct class. The remarks under Class IV as to analytical data apply (see page 134).

The "Foot" Oils.

These consist of the oils obtained from the hoofs of the sheep, horse, and cattle (Neat's foot). They differ only slightly in their characters and reactions, and a description of the last named will serve for the three.

§ 137. Typical Oil. NEAT'S FOOT OIL.

I. General and Analytical.

Character.

Pale yellow in COLOUR.

Almost ODOURLESS.

A small quantity of STEARINE separates.

Physical and Chemical Data.

	Test	Average figure	Usual variations	Extremes recorded
(1)	Specific gravity at 15·5° C.	·917	·9165—·9175	
(3)	Solidifying point fatty acids ° C.		26°—26·5°	
(4)	Refractive index 40° C. (Z. B.)	64 at 20°		
(5)	Viscosity 15·5° C. ($\eta \times 100$)	108	104—113	
(8)	Iodine value	70	69—72	
	(*a*) Bromine thermal test	12·5		
	(*b*) Maumené	47	43—49	
(9)	Elaidin	hard		
(10)	Saponification value	195	194—197	
(14)	Acid value	5	4—10	
(15)	Unsaponifiable matter per cent.	—	·1—·6	

Special tests.

None.

Chemical Composition.

About 78 per cent. Oleic acid, together with Stearic and Palmitic. Contains apparently *no Linolic acid.*

Adulteration.

(1) Vegetable, (2) Marine and (3) Mineral Oils, may be added.

Best detected by

(1) Iodine value of liquid fatty acids: Phytosteryl acetate test.

(2) Insoluble bromides (melting point to confirm).

(3) Unsaponifiable matter.

II. Technical.

Source.

The hoofs of cattle.

Method of obtaining Oil.

The cleaned hoofs are boiled with water in steam-jacketed pans, and the oil, as it is freed, is skimmed off the surface.

It is then settled, washed and filtered.

Employed largely as a soap emulsion for treating hides in the leather industry. Also for lubricating fine machinery.

§ 138. LARD OIL.

Character.

Almost colourless oil, with faint odour of lard.

Physical and Chemical Data.

	Test	Average figure	Usual variations	Extremes recorded
(1)	Specific gravity at 15·5° C.	·9155	915—916	
(4)	Refractive index 40° C. (Z. B.)	41		
(8)	Iodine value	85	67—88	
(9)	Elaidin	hard		
(14)	Acid value	small		
(15)	Unsaponifiable matter per cent.	0·4	·3—·5	

Chemical Composition.

19—25 per cent. solid acids; rest Oleic with little Linolic acid.

Uses.

High grades for edible products, cf. margarine, "compound lards."

Lower grades. Lubricating, burning, wool treating.

§ 139. Solid Glycerides.

The fats (many of which are oils at the average temperature of the countries where they are produced) vary in consistence according to the percentage of **liquid glycerides** which they contain. As the consistence is little guide to the properties of the fats, it seemed to the authors a more rational method to group them according to the **mean molecular weight of the fatty acids**. This method has the further advantage that it is unaffected by the artificial process of obtaining a further hardening of the fat by means of the **hydrogenation** of the liquid glycerides. As the mean molecular weight is proportional to the **saponification value**, the authors propose the following subdivision based on this determination:

Vegetable fats (Class VI).

- **Group 1.** Saponification value below 200
- **Group 2.** „ „ between 200 and 225
- **Group 3.** „ „ above 225.

Animal fats (Class VII).

- **Group 1.** Saponification value below 200
  Body fats.
- **Group 2.** Saponification value above 200
  Milk fats.

The following analytical data are *unimportant* in the case of the vegetable and animal fats:

- **(5)** Viscosity.
- **(7)** Polarimeter (except Chinese Vegetable tallow, q.v.).
- **(9)** Elaidin test.
- **(11)** Insoluble Bromide value.
- **(13)** Acetyl value.

The Solubility (Valenta) **(6)** is of special value in the case of Coconut, Palm Kernel, and Butter fat, and the Reichert-Meissl and Polenske values **(12)** and **(12** *a***)** have reference only to these same fats.

THE VEGETABLE FATS

Group 1.	S. V. BELOW	200.	*Typical fat*,	Cacao-butter
,, **2.**	S. V. ,,	200—225.	*Typical fat*,	Palm oil
,, **3.**	S. V. ABOVE	225.	*Typical fat*,	Coconut oil

(2) SOLIDS

CHAPTER XI

CLASS VI. VEGETABLE FATS

§ 140. Group 1. Saponification value below 200.

TYPICAL FAT—CACAO BUTTER.

The solid glycerides of these fats contain mostly only small quantities of fatty acids of lower molecular weight than **stearic** (18 carbon atoms) and little or no volatile acids (4—12 carbon atoms).

The following are described:

			Average
[1] Shea butter	saponification value		187
[1] Mowrah butter	"	"	192
Illipé butter	"	"	189
Cacao butter	"	"	194
Borneo tallow	"	"	195

§ 141. Shea Butter.

A fat, chiefly from West Africa, with usually a greyish colour, and strong smell. Refined, the fat is white and almost odourless, and is used for edible purposes. It is crudely prepared by boiling the ground kernels with water. Stearic acid, about one-third of total fatty acids[2].

Titer, 54°.
Saponification value, 179–184.
Iodine value, 54–63.

It is also characterised by a high proportion of unsaponifiable matter, 4–10 per cent.

§ 142. Mowrah Butter (Latifolia fat[3]).

From seeds of *Bassia Latifolia*, an Indian tree. The seeds contain over 50 per cent. of fat. The best qualities have a pleasant aromatic odour, and a yellow colour which is readily bleached by chemical agents.

Saponification value, 192.
Iodine value, 62.

Stearic acid is present to the extent of 13–25 per cent.[4]. It is used as an edible fat in India, and is employed in the soap and candle industries in Europe.

[1] These fats, and certain lesser known ones, of saponification values below 200, may contain proportions of palmitic and of myristic acids. In this case, their high mean molecular weight probably results from their also containing arachidic acid, or other acids of higher molecular weight than stearic. In the case of Shea Butter the unsaponifiable matter also lowers the s. v.

[2] Southcombe. *Jour. Soc. Chem. Ind.* 1909, 499.

[3] Bolton and Revis. [4] Menon.

§ 143. Illipé Butter (Longifolia fat[1]).

A fat similar in appearance to Mowrah, and is employed in the same manner. The two are often confounded.

Saponification value, 188. Iodine value, 50–60.

Stearic acid, 12–20 per cent.[2]

§ 144. TYPICAL FAT. CACAO-BUTTER (COCOA-BUTTER).

I. General and Analytical.

Character.

COLOUR—yellowish. ODOUR—strongly recalls cocoa.

CONSISTENCY—hard and brittle.

Physical and Chemical Data.

	Test	Average figure	Usual variations	Extremes recorded
(1)	Specific gravity	**·970**	964—976	
(2)	Melting point °C. complete fusion	**33**	32—34	28[3]—
(3)	Titer °C.	**49·5**	48—50	
(4)	Refractive index 40° C. (Z. B.)	**46·7**	46—47·5	
(6)	Solubility. True Valenta	94·0°[4]		
	$\frac{V \times 10}{80}$	11·7[4]		
	Alcohol reagent	76·0°[4]		
(8)	Iodine value	**37**	35—38	32·8[5]—41[5]
(10)	Saponification value	**193**	190—195	202[6]
(14)	Acid value	**0·5—1·5**		
(15)	Unsaponifiable matter	**0·8—1·2** per cent.		

Special Tests. Björklund's test (see practical part).

Chemical Composition.

Stearic acid 40 p.c.[7] About 30 p.c. of Oleic acid, and Linolic acid is probably also present. Palmitic and Arachidic acids also occur.

Adulteration.

Coconut and Palmnut stearines—higher saponification value.

Vegetable oils—lower Titer, higher Iodine value.

Beeswax and Paraffin wax—increased unsaponifiable.

Tallow—Cholesteryl acetate test: Björklund test.

II. Technical.

Source.

From Cocoa beans, the product of the *Theobroma cacao* from the West Indies, Central America, etc.

[1] Bolton and Revis. [2] Menon. [3] Lewkowitsch.
[4] Fryer and Weston. [5] Strohl. [6] Filsinger.
[7] Hehner and Mitchell.

Content of Fat in Seed.

40—50 per cent. (whole seed).

Method of obtaining fat.

The beans are first "fermented" and then sun dried. (More modern drying plant is being installed.) They are then shipped to Europe. The further treatment takes place in the cocoa works. The beans are roasted, shelled, ground, and the fat expressed hot in hydraulic presses, usually after the addition of carbonate or bicarbonate of soda. Soaps will thus be formed if fatty acids are present, and these will remain in the fat.

Refinement.

The fat is not treated further as the odour and flavour of the cocoa is required to be retained.

Uses.

Used largely in confectionery and chocolate manufacture, in pharmacy, and as an absorbent medium in perfumery.

§ 145. Borneo Tallow.

Obtained from the seeds of many varieties of Dipterocarpus, natives of Borneo. The fat obtained by native methods has disagreeable characteristics; that prepared in Europe resembles cacao butter in odour and brittleness. It contains 66 per cent. of stearic acid in the mixed fatty acids[1]. Stated to be used as a substitute for cacao butter.

Saponification value, 195
Iodine value, 29–38
Melting point (complete fusion) 38°
} average figures.

§ 146. Group 2. Fats with saponification value between 200 and 225.

TYPICAL FAT—PALM OIL.

These fats contain notable amounts of fatty acids of lower molecular weight than stearic acid. Palmitic acid forms a large percentage of some, and stearic acid is almost absent. Little or no **volatile** fatty acids are present.

The following are described:

Chinese vegetable tallow	203	average sapon. value
Palm oil	201	„ „
Japan wax	219	„ „
Myrtle wax	212	„ „
[Nutmeg butter	150—190][2]	„ „

[1] Geitel.

[2] The saponification value is low on account of essential oil present. The fat evidently belongs to this group.

§ 147. CHINESE VEGETABLE TALLOW.

I. General and Analytical.

Character.

Pale to dark green COLOUR. Tallow ODOUR. Hard and brittle.

Physical and Chemical Data.

	Test	Average figure	Usual variations	Extremes recorded
(1)	Specific gravity 15·5° C.	**·916**	915—918	
(2)	Melting point (complete fusion)	**60°**	55°—65°	33[1]—
(3)	Titer	**52·5**	52—53·5	40[2]—56[3]
(4)	Refractive index 40° C. (Z. B.)	**44**	42—45	
(6)	Solubility. True Valenta	74·5°[4]		
	$\frac{V \times 10}{80}$	9·4[4]		
	Alcohol reagent	65·5°[4]		
(7)	Polarimeter	**0**		
(8)	Iodine value	**20**	18—32	—38[3]
(10)	Saponification value	**203**	199—206	179[5]—231[6]
(12)	Reichert-Meissl value	**0·7**		
(14)	Acid value	**7**	not usually over 20	
(15)	Unsaponifiable matter	**1·25**	1—2	

Remarks.

Many figures have been obtained by observers from tallow containing apparently varying amounts of *Stillingia oil* from the kernels. This is best detected by the Polarimeter (**7**), Stillingia oil giving $-6° \ 45'$ in a 200 mm. tube.

Chemical Composition.

No Stearic acid[7]. Contains apparently PALMITIC and OLEIC acids only. Probably mainly OLEO DIPALMITIN[8].

Adulteration.

Not usually practised, but should be free from STILLINGIA OIL, which is contained in the kernel. Recognised by optical activity, and a high refractive index.

II. Technical.

Source.

The tallow coats the seeds of the *Stillingia sebifera* a native of China, and cultivated in N. India.

Method of obtaining fat.

As the *kernels* contain Stillingia oil, this is either separately obtained, or the whole seed is crushed yielding a softer mixture than the true tallow; and known by the native name of *mou-iéou*.

The tallow is detached from the seeds by rollers or melted by steaming.

Uses. Mainly for candles and soap.

[1] Seifert. [2] Hohein. [3] Jules Jean. [4] Fryer and Weston.
[5] De Negri and Fabris. [6] Zay and Musciacco.
[7] Hehner and Mitchell, *Analyst*, 1896, 328.
[8] Klimont, *Monatsch. f. Chem.* 1903, 408.

§ 148. Typical Oil. PALM OIL.

I. General and Analytical.

Character.

Colour—variable.

Intermediate shades are obtained between orange (Lagos oil) and reddish (Congo oil). Bleaches rapidly on exposure to air and light.

Odour—very characteristic.

Consistency—variable according to acid value.

Physical and Chemical Data.

	Test	Average figure	Usual variations	Extremes recorded
(1)	Specific gravity 15·5° C.	**·922**	921—924	
(2)	Melting point (complete fusion)	**41°**	35°—43°	27[1]—
(3)	Titer	**44°**	36°—45°	
(4)	Refractive index 40° C. (Z. B.)	**42**	41—45	
(6)	Solubility. True Valenta	93·5°[2]		
	$\frac{V \times 10}{80}$	11·7[2]		
	Alcohol reagent	68·0[2]		
(8)	Iodine value	**54**	52—56	
(10)	Saponification value	**201**	200—203	196[3]—
(11)	Reichert-Meissl value	**1·0**	—	
(14)	Acid value	**20—185**		

Remarks.

The odour and colour of Palm oil are so characteristic that the colour tests proposed for its recognition seem superfluous.

Chemical Composition.

Nördlinger[4] gives 98 per cent. Palmitic, in solid acids. Liquid acids: Oleic and probably small quantity of Linolic acids.

Adulteration.

Not commonly adulterated. Water and sand are however often present, and an allowance is usually made if together they exceed 2 °/₀.

II. Technical.

Source.

Various species of the *Elœsis* palm, which is practically[5] confined to West Africa. The fruits are borne on heads, and consist each of soft pulp with an inner kernel (see Palm Kernel Oil, § 152).

Content of Oil. 55 to 65 per cent. of pulp.

[1] Fendler. [2] Fryer and Weston. [3] Moore.
[4] *Zeitschr. f. angew. Chem.* 1892, 110.
[5] Introduced into S. America and Philippines, but industry small.

Method of obtaining oil.

The ripe fruit is cut off in bunches from which it is picked and stored in the ground to undergo fermentation. It is thereby softened and in this condition is bruised and roughly strained yielding oil. The pulp is afterwards boiled with water and the oil skimmed off. Owing to these crude methods, probably upwards of two thirds of the oil is lost. The fresh fruits are hand-pressed for a native edible oil, which is not exported.

More modern methods are now obtaining owing to the establishment of European plant and procedure in West Africa.

Refinement.

The oil arrives in this country with a high percentage ot free fatty acids. This is due to the hydrolytic enzyme contained in the pulp. Decomposition still proceeds on storing the oil until hydrolysis is almost complete. Refinement consists chiefly in removing the colour. This is accomplished by:

(1) Blowing AIR through the oil heated to 100°—150° C.,

(2) Treatment with OZONE (colour liable to "revert"),

(3) Bleaching with DICHROMATE OF SODA.

(1) **Air bleaching.** The oil, owing to its high acidity, is preferably treated in a lead-lined timber tank. It is heated by means of a copper coil, through which may be passed high pressure or superheated steam as required. When the temperature has reached 100° C. the air blowing is commenced and has the effect of agitating the oil. The temperature is then raised until bleaching is complete. If necessary superheated steam must be employed, but with many kinds of oil saturated steam, at a pressure of 100 to 150 lbs. per square inch suffices.

(2) **Bleaching with Dichromate of Soda.**

The oil is treated in a tank similar to the above, and provided with an "open" steam coil. It is raised with steam heat to a temperature not exceeding 60° C., and the mixed Dichromate and Hydrochloric acid (1 to 3 per cent. of the oil) with about 10—15 parts of water run slowly into the oil which is meanwhile kept agitated by means of compressed air. When the bleaching is finished, as judged by taking small samples of the bulk, the chrome liquor is allowed to settle out, and is run off. The oil is then washed with weak acid, and with water alone to free it from all traces of chromium and mineral acid.

Properties and Uses.

"Lagos oil" represents a high grade of oil, and "Congo oil" a typical low grade.

Used in large quantities for hard soap (bleached and unbleached) and for candles (bleached and deglycerinated). Also unbleached for the protection of the plate from oxidation in the tin-plate industry.

Used in railway wagon greases it gives the characteristic yellowish colour of the lubricant.

§ 149. JAPAN WAX.

I. General and Analytical.

Character.

Pale yellow or light brown in COLOUR.

ODOUR pronounced and characteristic.

CONSISTENCY hard and brittle with conchoidal fracture. Resembles bleached beeswax.

Physical and Chemical Data.

	Test	Average figure	Usual variations	Extremes recorded
(1)	Specific gravity at 15·5° C.	·987	·975—1·000	
(2)	Melting point (complete fusion)	52°	51°—53°	
(3)	Solidifying point fatty acids	59°	58°—60°	
(4)	Refractive index 40° C. (Z. B.)	48	47·6—49·7	
(6)	Solubility. True Valenta $\frac{V \times 10}{80}$ Alcohol reagent			
(8)	Iodine value	6	4·5—15	
(10)	Saponification value	220	217—222	
(14)	Acid value	6	usually not over 20	
(15)	Unsaponifiable matter	1·0	·7—1·5	

Soluble boiling alcohol; insoluble cold alcohol.

The term "wax" is of course a misnomer (see page 7), as it consists entirely of glycerides.

Chemical Composition.

Almost entirely composed of Palmitin and free Palmitic acid, together with a small proportion (less than 1 %) of DIBASIC ACIDS (e.g. Japanic acid), and probably some soluble acids.

Adulteration.

Occasionally starch is said to be mixed with the "wax," and also water.

Tallow might be employed as adulterant, and would be indicated by high Iodine Value and lower melting point. True WAXES distinguished by unsaponifiable matter and percentage yield of Glycerol on saponification. Perilla oil (*v. infra*) will increase the Iodine Value.

II. Technical.

Source.

From the berries of various species of the Sumach tree (*Rhus.*) cultivated in Japan and China for the lacquer it yields (the wax being a by product).

Content of fat.

15 to 25 per cent. of berry.

Method of preparing fat.

The wax occurs as a greenish coating to the berries. These are gathered and stored until "ripe," and are then crushed in a rough trough, and the kernels of the berries separated. The remaining crude wax is pressed in wedge presses. *Perilla* oil is stated to be used to increase the yield of wax.

Refinement.

By remelting and sun bleaching the wax, which is meanwhile kept moist. The bleached wax is exported in the form of slabs.

Properties and Uses.

The wax "effloresces," a fine crystalline deposit being formed on the surface.

Used in floor polishes and in the leather industry.

§ 150. Myrtle Wax.

Myrtle 'wax' is another instance of a true fat mis-named. Like Japan wax, it consists chiefly of Palmitin. It is obtained from the berries of a shrub (Myrica) growing on the Atlantic Coast of America and in South Africa. The wax is obtained by boiling out with water. It is light green in colour, and fairly hard and brittle, but melts at the comparatively low temperature of 40°—44°C. Used as a candle material and makes a fair polish for leather etc. when dissolved in turpentine.

§ 151. Group 3. Vegetable fats with saponification value over 225.

TYPICAL FAT—COCONUT.

These fats are characterised by the presence of glycerides of fatty acids of very low molecular weight. These are caproic (6 carbon atoms), caprylic (8 C. atoms), capric (10 C. atoms) and usually lauric (12 C. atoms). There is however **no butyric acid** present (see page 37).

The fats consequently give high figures in the Reichert-Meissl and in the saponification values.

There are several lesser known fats belonging to this group, but the only members which are important technically are:

{ Palm kernel oil
{ Coconut oil.

The Valenta[1] test is here of importance.

[1] See footnote on page 75.

§ 152. PALM KERNEL OIL.

I. General and Analytical.

Character.

Pale yellow to white in COLOUR.

ODOUR and FLAVOUR characteristic and "nutty"—low grades are sharp and unpleasant. CONSISTENCY salve-like.

Physical and Chemical Data.

	Test	Average figure	Usual variations	Extremes recorded
(1)	Specific gravity 15·5° C.	—		
(2)	Melting point (complete fusion)	**25°**	23°—29°	
(3)	Solidifying point fatty acids	**23°**	20°—25°	
(4)	Refractive index 40° C. (Z. B.)	**36·5**	36—38	
(6)	Solubility. True Valenta	24·0°[1]		
	$\frac{V \times 10}{80}$	3·0[1]		
	Alcohol reagent	40·0°[1]		
(8)	Iodine value	**13·5**	10—17	
(10)	Saponification value	**246**	244—248	
(12)	Reichert-Meissl value	**5·2**	5—6·8	
	(*a*) Polenske	**9·8**	10—12	
(14)	Acid value	**5—22**		
(15)	Unsaponifiable matter	Under ½ per cent.		

Special Tests.

Kirschner Value 1·07[2] (see page 173).

Baryta Value (soluble) 32·9[3] (insoluble) 303·7[3].

Chemical Composition[4]. Percentage of fatty acids present in combination.

Caproic acid	2	Lauric acid	55	Stearic acid	7
Caprylic ,,	5	Myristic ,,	12	Oleic ,,	4
Capric ,,	6	Palmitic ,,	9		

Adulteration.

Not usually adulterated.

II. Technical.

Source.

From the kernels of the Elœsis palm (page 154) of West Africa.

Content of oil.

The kernels contain 44—49 per cent. of oil.

Method of obtaining oil.

Native labour is almost exclusively employed for collecting the kernels and cracking the shells. Very large quantities are wasted. The kernels

[1] Fryer and Weston. [2] Bolton and Revis.
[3] Avé Lallement. [4] Elsdon, *Analyst*, 1914, 78.

are shipped to European ports—chiefly Hamburg (Marseilles during the war).

The oil is either expressed or extracted from the kernels.

The expression process consists in screening the kernels, grinding between rollers (see Plate XIII), and pressing the pulpy material twice at about 50° and 60° C. respectively (see page 213, cage press).

The press cakes are used for cattle feeding, and contain only about 7 per cent. of oil.

At Hamburg extraction with Carbon Disulphide is carried on. The residue is only fit for manure.

Refinement.

This is necessary for edible fat, and is accomplished by the use of soda or potash which combines with free fatty acids and carries down impurities and colouring matters. The residue is sold for soap stock.

Properties and uses.

As it closely resembles Coconut Oil it may be used for the same purposes, e.g. cold process soaps, "washer" soaps, "marine" soaps, edible fats. The oil is also pressed yielding a "stearine" used as a "chocolate fat."

§ 153. Typical Oil. COCONUT OIL.

I. General and Analytical.

Character.

Pure white to brownish COLOUR.

ODOUR AND FLAVOUR. Characteristic sweet "nutty." Low grades pungent and acrid.

CONSISTENCE. Somewhat brittle in winter; salve-like in summer.

Physical and Chemical Data.

	Test	Average figure	Usual variations	Extremes recorded
(1)	Specific gravity 15·5° C.	**·9258**	·9255—·9265	
(2)	Melting point (complete fusion)	**24·5°**	23°—26°	
(3)	Solidifying point fatty acids	**23°**	21°—25°	
(4)	Refractive index 40° C. (Z. B.)	**35·5**	35—36	
(6)	Solubility. True Valenta	12·0°[1]		
	$\frac{V \times 10}{80}$	1·6[1]		
	Alcohol reagent	34·0°[1]		
(8)	Iodine value	**8·5**	8—9	
(10)	Saponification value	**257**	255—260	
(12)	Reichert-Meissl value	**7·5**	6·5—8	
	(*a*) Polenske value	**16·5**	15—18	
(14)	Acid value	**10**	6—14 (good varieties)	
(15)	Unsaponifiable matter	**0·2**	0·15—3 (low)	

[1] Fryer and Weston.

Special tests.

Kirschner 1·6 to 1·9.

[1]Baryta values (*a*) Insoluble 294·5
(*b*) Soluble 57·3
$a-(200+b)=+37{\cdot}2.$

Chemical Composition.

Elsdon[2] found percentage of fatty acids in glyceride as follows:

	Per cent.		Per cent.
Caproic acid	2	Myristic	20
Caprylic ,,	9	Palmitic	7
Capric ,,	10	Stearic	5
Lauric ,,	45	Oleic	2

(Stearic probably too high and Oleic too low.)

Adulteration.

Seldom practised. Readily discovered by the usual constants except in the case of PALM KERNEL OIL. The "Baryta value"[3] should prove useful in the case of large admixtures of this oil.

II. Technical.

Source.

The fruit of the coconut tree (*Cocos nucifera* and *butyracea*).

The chief productive areas are Ceylon, South America, India, Philippines, South Sea, Cochin China.

Content of Oil.

Fresh kernels 30—40 per cent.
Dried copra 50—74 ,, ,,

Method of Obtaining Oil.

The native methods of preparing the oil consist in exposing the sliced kernels to the sun, and collecting the oil which runs off. A better quality is obtained by pounding the kernels and boiling out the fat with water, one of the chief methods of obtaining the COCHIN OIL. Roughly improvised presses are also in use. The oil is carried by natives to the coast towns.

A very large amount of dried kernel is exported under the name of COPRA.

The husk is removed, and the split nut either dried by exposure to sunlight, or in fire-heated kilns. The latter gives an inferior

[1] Avé Lallement. [2] *Analyst*, 1914, 78.
[3] A method has been proposed by Burnett and Revis (*Analyst*, 1913, **38**, 255) to distinguish between these oils by observation of the "turbidity temperature" of the Barium salts of the insoluble volatile acids in alcoholic solution (of definite strength).

product, except where hot air is employed as a drying agent, when a fine white copra results.

The copra imported into Europe is pressed for oil in a similar manner to Palm kernels. The pulp is twice expressed at moderate temperatures (55° to 60° C.) in a press of the "cage" variety (see page 213).

Carefully treated, copra yields a product which is worked up entirely for edible purposes.

The cakes are used for cattle feeding and contain up to 10 per cent. of oil.

Extraction of the oil with volatile solvents is unremunerative owing to the low value of the extracted residue.

Refinement.

The oil is refined in a similar manner to that employed in the case of Palm Kernel oil (page 159).

Properties and Uses.

Three grades are distinguished in commerce:

Cochin.

Originally obtained from Malabar, but the name is now a generic one for oil produced from fresh kernels by careful methods. As imported is white and of good sweet odour and flavour. The percentage of free fatty acids is low.

Ceylon.

This usually represents a lower quality of imported oil, though the best varieties are scarcely distinguishable from "Cochin." It varies from yellowish brown to white. The odour is less pleasant than cochin, being often somewhat sharp. The amount of free fatty acid is very variable.

Copra.

The term is used commercially to designate the lowest quality of oil. In point of fact, however, the quality is very variable according to the care employed in the preparation of the dried kernels. The poorest varieties of the oil are brown in colour, of a very rank odour, and contain a large percentage of free fatty acids.

Coconut oil is used in enormous quantities in edible fats. Many trade preparations, such as "nucoline," etc., consist largely if not entirely of coconut oil. It forms an important ingredient of margarine, and is also sold as "nut butter" (emulsified with milk).

It has also a very extended use in the hard soap industry. The use of coconut oil is almost a necessity in milled and "washer" soaps owing to the freely lathering properties which it imparts to them.

Like Palm kernel oil it can be used in the preparation of sea water soaps (marine soaps) which are soluble in dilute brine. Owing to the ease with which it saponifies, it is a necessary ingredient in "cold process soaps," its function being to commence the process of saponification.

THE ANIMAL FATS

Group 1. S V. BELOW 200.
Body fats. *Typical fat*, Tallow

" **2.** S. V. ABOVE 200.
Milk fats. *Typical fat*, Butter fat

CHAPTER XII

CLASS VII—ANIMAL FATS

§ 154. Group 1. Saponification value below 200.

TYPICAL FAT—TALLOW.

These are all contained in the fatty or adipose tissue of various animals—mostly mammals. They consist mainly of the glycerides of **stearic** and **oleic** acids, with small amounts only of linolic acid.

Fats of fish-eating animals, such as the wild goose, ice bear and others show distinct drying properties, and give high iodine values. This is almost certainly due to the absorption of the unsaturated fatty acids of fish oil into the body fat of these animals.

An important question is raised by the consideration of the influence of the food of the animal on these fats. It has been proved that unsaturated fats occurring in the food are in process of time absorbed to some degree by the body fat not only in the case of fish-eating animals as above, but also with domesticated beasts. Thus lard of hogs fed on cotton cake shows a higher iodine absorption value due to an increased percentage of linolic acid[1].

Researches are in progress which appear to show that glycerides are hydrolysed in the body, the fatty acids becoming absorbed and being subsequently re-esterified by cell activity.

Certain other constituents of fats are also absorbed and appear in the body fat. **Thus the substance which produces the Halphen reaction in cotton seed oil appears in the lard oil of hogs fed on cotton cake.**

In these circumstances it thus becomes a matter of great moment whether the distinguishing alcohol—phytosterol—of vegetable oils is also absorbed and contained in the body fat of animals fed on products containing vegetable oils. A number of researches by different observers[2] have proved conclusively that **phytosterol is not absorbed by the fat of the animal in these conditions.**

The distinction between animal and vegetable fats by the identification of cholesterol and phytosterol holds good thus far, irrespective in the case of animals of the nature of the diet.

[1] See also effect on butter fat by feeding cows with ground nut and cottoncake. Cranfield, *Analyst*, 1916, **488**, 336–339.

[2] König and Schluckebier, *Zeit. f. Unters. d. Nahrgs. u. Genussm.* 1908 (xv), 642.

§ 155. TYPICAL FAT. **LARD.**

I. General and Analytical.

Character.

White, translucent and of salve-like CONSISTENCY. Finest qualities are usually **granular.** ODOUR and FLAVOUR vary according to quality.

Physical and Chemical Data.

	Test	Average figure	Usual variations	Extremes recorded
(1)	Specific gravity 15·5° C.	**·936**	·934—·938	
(2)	Melting point (complete fusion)	**42°**	36°—48°	
(3)	Solidifying point fatty acids	**40°**	36°—42°	
(4)	Refractive index 40° C. (Z. B.)	**50**	48—53	44·8[1]—57·3[2]
(6)	Solubility. True Valenta	87·0°[3]		
	$\frac{V \times 10}{80}$	10·9[3]		
	Alcohol reagent	72·5°[3]		
(8)	Iodine value	**62**	56—66	49·9[4]—85[2]
(10)	Saponification value	**196**	194—199	
(12)	Reichert-Meissl value	**0·6**	0·5—0·7	
(14)	Acid value	**0·8**	·2—1·5	
(15)	Unsaponifiable matter	**0·35**	0·2—0·5	

Remarks.

Within the above limits the constants vary in fat taken from different parts of the body. The Iodine value of the liquid fatty acids is important in the recognition of adulteration. It is from 92—115.

Chemical Composition.

The fatty acids of lard consist of about 40 per cent. solid and 60 per cent. liquid acids. The latter consist of oleic and linolic in about the proportion of 5 to 1. The solid acids[5] consist of myristic, lauric, stearic and palmitic in order of importance. The proportions vary in the fat from different parts of the animal.

Adulteration.

Lard is very liable to adulteration. The chief and likely adulterants are as under:

Water. Determined by drying a weighed quantity at 105° C.

Vegetable Oils.

COTTON and SESAMÉ oils are recognised by the Halphen and Baudouin colour reactions, except in the case of Hogs fed on cake

[1] Dennstedt. [2] Farnsteiner. [3] Fryer and Weston. [4] Dieterich.

[5] Bömer (*Analyst*, 1913, **38**, 204) has isolated palmito-distearin from lard, and gives evidence to show that it is the α compound, see p. 14.

from these oils. In this case the iodine values of the oil (and of the liquid fatty acids in case of cotton) will probably be higher than normal. For proof of the presence of these oils in small amount, recourse must be had to the phytosteryl acetate test (see practical part)[1]. The sulphur chloride test is useful.

ARACHIS OIL. Renard's test. The melting point of the arachidic acid should be obtained.

MAIZE OIL. Solidifying point of fatty acids (if cotton be absent).

Animal Fats.

TALLOW AND BEEF STEARINE.

This is a very difficult problem if the amount of adulteration fall below 10 per cent. The most important test is the solidifying point of the mixed fatty acids, together with a microscopical examination of the crystals formed from an ethereal solution of the lard.

Pure lard crystals are oblong plates with chisel-shaped ends (the loin fat crystallises best: back fat difficulty crystallizable). See practical part.

Beef crystals appear as curved tufts of very fine needles. See plate in practical section.

The Stock-Belfield method for detecting beef stearine is as follows:

Two standard sets of mixtures are prepared.

(A) Lard (pure) of melting point
34°—35° C. with 5, 10, 15, and 20 per cent. of Beef stearine. M.P. 56° C.

(B) Lard (pure) of melting point
39°—40° C. with 5, 10, 15, and 20 per cent. of Beef stearine. M.P. 50° C.

(1) Melting point (capillary method) obtained of lard *under examination*. It is then compared with (A) or (B) according to which of the two it is nearest in melting point.

(2) 3 c.c. of the melted fat is dissolved in 21 c.c. of ether in a graduated cylinder of 25 c.c. capacity.

(3) 3 c.c. of the set of mixtures are similarly dissolved.

(4) The cylinders are cooled to 13° C. for 24 hours.

(5) The amount of beef stearine is approximately indicated by comparison of the crystals formed in (2) with those formed in the sets of mixtures.

This is confirmed by the examination microscopically for beef stearine crystals.

II. Technical.

Source.

The fat from all parts of the Pig (formerly it signified the LEAF FAT only).

Preparation.

The large American packing houses employ a continuous method of treatment for the working up of the various products of the animal.

[1] Paraffin wax is sometimes purposely added to render this test ineffective.

Immediately after slaughtering, the leaf fat (that which surrounds kidney and bowels) is removed from the hog, cooled, and "rendered" (see page 210) in jacketted vessels at a temperature not exceeding 50° C. This yields

"NEUTRAL LARD NO. 1."

The fat from the back is treated similarly and gives the

"NEUTRAL LARD NO. 2."

The residue from "Neutral lard No. 1" is digested with high-pressure steam and yields

LEAF LARD.

Residues from "Neutral lard No. 2" similarly treated give

CHOICE KETTLE-RENDERED LARD.

Trimmings, and fatty tissue from other parts of the animal, digested with high-pressure steam give

PRIME STEAM LARD.

The intestines and viscera are worked up in digesters and yield GREASES of various qualities, as also do diseased hogs, which are treated by being boiled in strong caustic alkali, which destroys all tissues, leaving only the fat.

Refining.

(1) Neutral lard is settled and "salted" (to remove traces of water and fibre) and is then ready for use.

(2) Other qualities are refined with Fuller's earth (see page 218) and rapidly cooled.

Properties and uses.

NEUTRAL LARD 1 AND 2 are used almost exclusively for margarine (oleomargarine) and confectionery.

The other lards are suitable for domestic purposes, and the GREASES are employed for soap-making (mainly high-class toilet and shaving soaps).

LARD STEARINE results on pressing the carefully crystallised fat. It is used as a "stiffener" for softer lards.

LARD OIL results from the above (see page 147).

§ 156. BONE FAT.

I. General and Analytical.

Character.

RENDERED.

White and almost odourless to brown and offensive.

EXTRACTED.

Dark colour and sharp offensive odour.

CONSISTENCE. (All grades.)

"Buttery" to "tallowy."

Physical and Chemical Data.

	Tests	Average figure	Usual variations	Extremes recorded
(1)	Specific gravity at 15·5° C.	**·915**	914—916	
(2)	Melting point (complete fusion)	**21°**	20°—22°	
(3)	Solidifying point fatty acids	**40°**	36°—42°	
(4)	Refractometer 40° C.			
(8)	Iodine value	**52**	50—55	46[1]—63[2]
(10)	Saponification value	**190**	187—195	172[2]—
(14)	Acid value	**1—50**	according to degree of freshness of bones	
(15)	Unsaponifiable matter	**1·0**	0·5—1·5	1·8[3]—

Remarks.

Bone fats normally contain lime salts, such as calcium stearate, oleate, etc., and also salts of the lower fatty acids as butyrate (and lactate). It is therefore necessary to obtain the content of fatty matter. This is determined as follows[4]:

(1) 10 grammes of sample are weighed in small conical flask.

(2) 3 to 5 drops strong HCl added.

(3) It is warmed on water bath for one hour with agitation (lime salts thereby being decomposed).

(4) 40 c.c. Petroleum ether added, and solution shaken.

(5) Filtered into tared flask; ether evaporated, and residual fat weighed.

DIRT AND GROSS IMPURITIES are determined by drying and weighing the residue from filtration of the ethereal solution.

LIME SALTS are estimated by incinerating say 10 grammes of the sample. The lime is obtained as carbonate.

WATER. This is best found by difference, as the water is tenaceously held by the lime salts and is difficult to drive off.

Chemical Composition.

This varies greatly, but may be taken as about 75 per cent. oleic, 15 per cent. stearic and probably some palmitic[5].

Adulteration.

Waste fats, recovered grease and wool grease. The latter are readily detected by the estimation of the unsaponifiable matter.

II. Technical.

Source.

Bones of all kinds in all conditions of freshness and of putridity.

Content of fat.

12—20 per cent.

[1] Wilson. [2] Troicky. [3] Shukoff and Schestakoff.

[4] Shukoff, *Chem. Revue*, 1898, 6.

[5] See Nerthing, *Biochem. Zeits.* 1908, 167.

Preparation.

Bone fat is the product of "degreasing" the bones prior to glue extraction, or distillation for obtaining animal charcoal and "distilled bone oil."

(1) Bones boiled in open vessels with water (now abandoned owing to nuisance created).

(2) Bones broken up in a mill (see Plate XIV) and treated with steam under pressure in autoclaves (steam pressure 30 to 40 lbs. per square inch). The bones are usually cleansed in a special washer before treatment. A prolonged treatment dissolves gluey constituents, and is undesirable.

(3) Bones extracted with volatile solvents. The usual solvents employed are Petroleum benzine, and Shale oil boiling between 100°—130° C. One of the continuous forms of extractors is employed such as "Thorne's," the solvent boiling during the extraction. It is essential that the boiling point should be above that of water, as by this means the bones become gradually dried, and the solvent is able to extract the fat completely.

Refinement.

Rendered fat is fairly easily bleached unless the acidity is very high. For this purpose DICHROMATE and hydrochloric acid are employed (page 222). The PEROXIDES have also been used (page 222).

Extracted fat is not usually bleached, but is distilled, a white fatty acid being obtained which is used as candle-material.

Properties and uses.

The extracted fat is almost entirely employed in soap-making. A certain amount is pressed yielding "Bone Oil"[1] which is a valuable lubricant.

The extracted fat is distilled (see above) and pressed yielding a STEARINE for candle-making and an OLEINE. These cannot be employed for soap-making owing to the high proportion of unsaponifiable matter they contain.

§ 157. TALLOW. (BEEF AND MUTTON FAT.)

I. General and Analytical.

Character.

COLOUR—white to dark brown.

ODOUR—faint to very disagreeable (mutton is more pronounced than beef).

FLAVOUR—according to grade. Mutton is less agreeable.

[1] This must not be confounded with distilled bone oil, which is a product of the destructive distillation of bones after "degreasing" and contains organic bases. It is a by-product in the production of "animal charcoal."

CONSISTENCE.

Varies according to the food, sex, and part of the body from which it is obtained. Mutton is normally harder than beef.

Physical and Chemical Data.

	Test	Average figure	Usual variations	Extremes recorded
(1)	Specific gravity at 15·5° C.	**·947**	·937—·953	
(2)	Melting point (complete fusion)	**48°**	47°—49°	
(3)	Solidifying point fatty acids	**44°** (B.) **45·5°** (M.)	43°—45° (B.) 45°—46° (M.)	
(4)	Refractive index 40° C. (Z. B.)	**48**	47—49	
(6)	Solubility. True Valenta	92·0°[1]		
	$\frac{V \times 10}{80}$	11·5[1]		
	Alcohol reagent	72·5°[1]		
(8)	Iodine value	**44** (B.) **41** (M.)	42—45 (B.) 38—42 (M)	
(10)	Saponification value	**195**	193—198	
(12)	Reichert-Meissl value	**0·5**		
(14)	Acid value	Varies according to grade. Best edibles not over 2.		
(15)	Unsaponifiable matter	**0·5**		

Chemical Composition.

It contains glycerides of Stearic, Palmitic and Oleic acids. Stearic acid 33–50 per cent. Oleic 50–60 per cent. Linolic and even Linolenic acid has been proved to exist in tallow, probably owing to the comparatively recent practice of feeding cattle and sheep on oil cakes. The glycerides have been definitely proved to be of a MIXED character, the following having been isolated: Stearodipalmitin, oleodipalmitin, oleopalmitostearin, palmitodistearin.

Adulteration.

Not common.

VEGETABLE OILS—detected by Iodine value and Phytosteryl Acetate Test. Also by melting points.

WHALE OIL—Insoluble Bromide value (negative with pure tallow) and odour of freshly liberated fatty acids.

PARAFFIN WAX and DISTILLED GREASES—unsaponifiable matter, and acid value.

II. Technical.

Source.

All parts of the body of the cow and sheep except the MILK.

[1] Fryer and Weston.

Preparation.

On the small scale—mainly as an adjunct to slaughter-houses—fat from all parts of the body is rendered by heating at low temperatures and straining from the adherent tissue.

In the large American packing-houses, the fat is at once removed from the slaughtered animal and the kidney and bowel fat kept separate. It is hardened by standing and immersion in ice-cold water, and is then shredded and rendered in melting pans at a temperature of about 40° C. Salt is sprinkled over to assist the separation of the tissue. The clear fat forms the "PREMIER JUS." It is allowed to crystallise and pressed, forming "OLEO OIL" and "OLEO-STEARINE."

The fat from the rest of the carcase is rendered at temperatures of 100° C. upwards and is termed "RENDERED TALLOW" (see page 210).

Refinement.

The "Premier Jus" is refined by re-melting and salting. Commercial and off-colour tallows are bleached by means of dichromate, manganese dioxide, and the peroxides (see page 220); ozone has also been employed. Dichromate bleached tallows frequently have a greenish shade due to chromium.

Properties and uses.

The "Premier Jus" and its products, Oleo oil (oleomargarine) and Oleo stearine yield white fats which are almost tasteless and odourless, and are mainly used in margarine and cooking preparations (lard compounds, "nut butters," etc.).

"Rendered Tallow" of all grades is employed in a great variety of ways, but chiefly for soap-making, candle-making (after deglycerination), and lubricating.

§ 158. Group 2. Milk fats.

TYPICAL MEMBER—BUTTER FAT.

The milk fats are obtained from the secretion of the mammary glands of the mammalia. Although the milk from various mammals varies somewhat in composition[1], the fat appears to be much more uniform. The only important member of the group, from a technical point of view, is cow butter fat. The characteristic acid is **butyric**, and other volatile acids are also present. The milk fats resemble the coconut group in the vegetable group, but the latter contain *no butyric acid.*

[1] We are not here concerned with a description of milk and butter, information on which may be obtained from the many treatises on the subiects.

§ 159. TYPICAL FAT. **BUTTER FAT.**

I. General and Analytical.

Character.

As obtained by melting and filtration from **butter**, it has an almost white to yellow colour, and retains its odour of butter. The consistence is a little harder than butter, and the fat is almost tasteless.

Physical and Chemical Data.

	Test		Average figure	Usual variations	Extremes recorded
(1)	Specific gravity at 15·5° C.		**0·938**	·936—·942	
(2)	Melting point (complete fusion)		**31°** C.	28°—34°	
(3)	Solidifying point fatty acids		**35°**	33°—37°	
(4)	Refractive index 40° C. (Z. B.)		**43·5**	42·5—44	
(6)	Solubility.	True Valenta	37·0°[1]		
		$\frac{V \times 10}{80}$	4·6[1]		
		Alcohol reagent	46·0°[1]		
(8)	Iodine value		**33**	26—38	—50[2]
(10)	Saponification value		**227**	220—232	
(12)	Reichert-Meissl value		**28**	25—30	17[3]—36·8[3]
	(*a*) Polenske		**2·3**	1·7—2·9	1·3[3]—3·5[3]
(14)	Acid value		Very small, except in fats from rancid butters		
(15)	Unsaponifiable matter		**0·35**	·3—·45	

Remarks.

The figures given by undoubtedly genuine samples of butter fat are liable to considerable variation. This is especially the case in regard to the percentage of volatile fatty acids. Thus the Reichert-Meissl values have an extreme variation of almost 100 per cent. (see above).

Special Tests.

Kirschner value—20 to 26 (see practical work).

(*a*) Insoluble Baryta value—usually below 260.

(*b*) Soluble „ „ — 55 to 67.

$a - (200 + b)$ is *negative*[4].

Chemical Composition.

Butyric acid occurs to the extent of 4·5 to 6 per cent. of the mixed fatty acids, caproic acid probably about 1 per cent., caprylic and capric acids together about 0·3 per cent. Amongst the higher saturated acids have been recognised lauric, myristic, palmitic,

[1] Fryer and Weston. [2] Van Rijn. [3] Polenske. [4] See pages 158, 160.

stearic and arachidic. Oleic is normally the only unsaturated acid present (except in small quantities), but linolic and even linolenic are said to have been found. This would undoubtedly be due to the employment of food containing glycerides of these acids.

As in the case of tallow, and probably many other oils and fats, the glycerides are—mainly at any rate—of a mixed character. OLEO-PALMITO-BUTYRIN has for instance been isolated.

The chemical composition varies somewhat widely in different samples of butter. The composition appears to depend on the breed of the cow, the mean temperature to which it is exposed, the kind and amount of its food, and the period of lactation.

Adulteration.

Owing to its high price and enormous consumption, no other oil or fat has been so subject to sophistication as butter. Further, the adulteration has kept, if not ahead, at all events well abreast of the analyst. Unfortunately adulteration of a systematic character is facilitated by the natural variations of pure butter itself.

The practice of analysts a few years ago was to base their conclusions on the REICHERT-MEISSL test, the amount of adulteration with lard, or oleomargarine being judged by the proportionate lowering of this figure. Although this method did not reveal adulteration of butters naturally rich in volatile acids, it gave on the average a fair indication. Since, however, coconut and palm kernel oils have been extensively employed for cheapening butter, this method breaks down, as it is possible to use up to 20 per cent. of these oils without their being detected by this test.

Polenske and others then showed that a reliable method of distinction could be based on the fact that coconut and palm kernel oils contain more *insoluble volatile acids* than butter fat. By filtering these acids off, and neutralising the alcoholic solution with $\frac{N}{10}$ alkali, a figure was obtained which he termed the "new butter value." Coconut gave values of 15 to 18, butter fat 1·6–3·5. Further, Polenske established that generally speaking, these values were proportional to the Reichert-Meissl values, and he constructed a table (see page 175) showing the approximate values for corresponding Reichert-Meissl figures.

A further elaboration was introduced by *Kirschner*, depending on the fact of the comparative solubility in water of silver butyrate, and the comparative insolubility of the caproate and caprylate (see page 37). By obtaining the silver salts and filtering off the insoluble ones, he obtains a measure of the butyric acid present in the sample[1].

Another method, due to *Avé-Lallement*, is based upon the extent of the solubility of the barium salts of fatty acids of butter and other fats. The results are calculated to barium oxide, and the figure for the soluble salts is increased by an arbitrary amount and subtracted from that of the insoluble figure so as to yield in the case of butter fat a *minus* quantity. Adulteration with 10 per cent. of coconut oil or of lard gives a *plus* figure in *both* cases. In this way it is presumed that adulteration with 10 per cent. of a mixture of lard and coconut oil can be detected, while neither the Reichert-Meissl nor the Polenske value would yield conclusive results.

In all cases where adulteration with vegetable oils or fats is suspected, it should be confirmed by the Phytosteryl acetate test which is the only *absolutely* reliable guide.

Further, great circumspection in the interpretation of results is necessary, and too great reliance must not be placed upon small observed differences from the figures hitherto obtained in the various tests as being conclusive evidence of adulteration.

II. Technical.

Source.

Cow's milk.

Content of fat.

Milk contains 3·0–4·5 per cent. of butter fat. The milk of Jersey cows is richer in fat than other breeds.

Preparation.

The anhydrous fat is rarely prepared.

Butter, obtained in the dairies by a process of churning, contains 8–16 per cent. of water, and curd (casein) with milk sugar (aggregating 1 to 2 per cent.).

For further details of manufacture the student is referred to works on dairy chemistry, etc.

Uses.

As stated above, almost entirely used as **butter**, the anhydrous fat being rarely prepared. In the case of "renovated butter" (so called) the fat is freed from the curd, etc. of stale material and emulsified with fresh milk. It is said to be detected by examination with the polariscope.

[1] Bolton, Richmond and Revis (*Analyst*, 1912, **37**, 183) have shown that the relationship between the Polenske and Kirschner values is very sensitive and constant. They claim by this means to detect the presence of as little as **1 per cent.** of butter fat in margarine in the presence of coconut oil, or palm kernel oil or both together. See also Cranfield (*Analyst*, 1915, 439).

Table of Reichert-Meissl values, and "Polenske" values corresponding to same. (Polenske.)

No.	Reichert-Meissl	Polenske	No.	Reichert-Meissl	Polenske
1.	19·9	1·35	18.	26·2	1·9
2.	21·1	1·4	19.	26·5	1·9
3.	22·5	1·5	20.	26·6	1·8
4.	23·3	1·6	21.	26·7	2·0
5.	23·4	1·5	22.	26·8	2·0
6.	23·6	1·7	23.	26·9	2·1
7.	24·5	1·6	24.	26·9	1·9
8.	24·7	1·7	25.	27·5	1·9
9.	24·8	1·7	26.	27·8	2·2
10.	24·8	1·6	27.	28·2	2·3
11.	25·0	1·8	28.	28·4	2·3
12.	25·1	1·6	29.	28·8	2·2
13	25·2	1·6	30.	28·8	2·5
14.	25·3	1·8	31.	29·4	2·6
15.	25·4	1·9	32.	29·6	2·8
16.	25·6	1·7	33.	29·5	2·5
17.	25·4	1·7	34.	30·1	3·0

LIQUID WAXES

THE SPERM OILS

B. NON-GLYCERIDES

(*a*) SAPONIFIABLE

I. LIQUID AND SAPONIFIABLE

This group comprises the LIQUID WAXES and all its members are obtained from marine mammals.

The most important analytical data are the Specific gravity (1), Viscosity (5), Saponification value (10), Unsaponifiable (15).

CHAPTER XIII

CLASS VIII. THE SPERM OILS

§ 160. TYPICAL MEMBER—**SPERM OIL.**

I. General and Analytical.

Character.

COLOUR—pale yellow to light brown.

ODOUR—odourless to distinctly "fishy."

"STEARINE" (spermaceti) deposited at ordinary temperatures, but "Winter Sperm" is free from solid wax.

Physical and Chemical Data.

	Test	Average figure	Usual variations	Extremes recorded
(1)	Specific gravity 15·5° C.	**·880**	·878—·883	
(3)	Solidifying point mixed fatty acids (+alcohols)	**11·5**		
(4)	Refractive index 40° C. (Z. B.)	**52**	50—54	
(5)	Viscosity ($\eta \times 100$) at 15·5° C.	**42**		
(6)	Solubility. True Valenta	84°[1]		
	$\frac{V \times 10}{80}$	10·5[1]		
	Alcohol reagent	68°[1]		
(8)	Iodine value	**86**	84—90	70·4[2]—96[3]
(10)	Saponification value	**124**	120—126	—150·3[4]
(11)	Insoluble bromide value	**2·5**	1·1—3·7	
(13)	Acetyl value	**5**	4·5—6·5	
(14)	Acid value	**1·5**	·5—2	
(15)	Unsaponifiable matter	**40**	39—42	

[1] Fryer and Weston. [2] Dunlop. [3] Bloxam. [4] Fendler.

Remarks.

Note especially low specific gravity (**1**), low saponification value (**10**), high unsaponifiable (**15**).

Chemical Composition.

Sperm oil consists mainly of an ester consisting of a monohydric alcohol and a fatty acid apparently of the oleic series. The exact nature of these is still unknown. It also contains varying proportions of the solid wax spermaceti (q.v.). No glycerides[1] are normally present.

Adulteration.

Frequently practised. The usual adulterant is a fatty oil, such as whale or seal oil. This is readily estimated by the amount of GLYCERINE yielded on saponifying[2]. The saponification value is also increased by the addition of a fatty oil. A judicious mixture of a mineral and fatty oil may however be used for adulteration without increasing the saponification value.

MINERAL OIL, with or without fatty oil, may be recognised by the acetyl value of the unsaponifiable matter.

Adulteration with mixtures of mineral and fatty oils may also be detected by the specific gravity, combined with the **Flash point** test.

II. Technical.

Source.

From the blubber and the head of the *Sperm* or *Cachalot* whale.

Yield.

Cows, 2⅓ to 3 tons ; Bulls, 7 to 9 tons.

Industry.

The oil from the blubber and head is usually kept separate, the latter, at first liquid, becomes thick on standing. The solid spermaceti is removed by refrigerating and pressing.

According to the lowness of the temperature employed, an oil of correspondingly high or low "cold test" is produced. The yield of oil is about ¾ of the original product. A further yield of cruder oil is obtained by a second pressing of the crude brown wax at a higher temperature.

The oil is usually bleached.

Uses.

Forms a valuable lubricant for light and high speed machinery as it has no "gumming" tendencies, and its viscosity does not fall off at high temperatures to the extent that other lubricating oils do. It is very extensively imitated with mineral distillates, the so-called "mineral sperms."

[1] See however Fendler, *Chem. Zeit.* 1905, 555.

[2] Since however genuine Sperm oils have apparently contained glycerides, this may not be taken as a quite final proof of adulteration.

§ 161. Arctic Sperm Oil—Bottlenose Oil.

Very closely simulates Sperm oil in all its properties, but is of less value on account of its greater tendency to "gum."

II. SOLID AND SAPONIFIABLE

§ 162. This group comprises the natural animal and vegetable waxes, and the mineral waxes known as "bitumen waxes," extracted from mineralized vegetable matter, such as lignite and brown coal. The bitumen waxes are distinguished from the natural waxes by containing no alcohols, the unsaponifiable constituents being hydrocarbons only.

THE NATURAL WAXES

Vegetable waxes. *Typical wax,* Carnaüba wax
Animal waxes. *Typical wax,* Beeswax

II. SOLID AND SAPONIFIABLE

(*a*) ALCOHOLIC

CHAPTER XIV

CLASS IX. THE NATURAL WAXES

§ 163. As previously stated, there is no chemical distinction possible, as in the oils and fats, between waxes from *animal* and from *vegetable* sources. The vegetable waxes are all of higher melting point, are more brittle, and have almost always a more or less pronounced aromatic odour on melting. The analytical data of importance are (1), (2), (4), (8), (10), (13), (14), (15).

§ 164. VEGETABLE WAXES.

These are a numerous class, but are mostly little known. They all occur as an excretion on the surface of the leaves and stems of plants.

The typical member is *Carnaüba wax*, and the only other wax important commercially is Candelilla wax.

§ 165. TYPICAL WAX. CARNAÜBA WAX.

I. General and Analytical.

Character.

Light yellow to dirty green (or bleached white).

ODOUR of *new mown hay* on melting.

Very hard and brittle.

Physical and Chemical Data.

	Test	Average figure	Usual variations	Extremes recorded
(1)	Specific gravity 15·5° C.	**·998**	·99—1·0	
(2)	Melting point ° C.	**84°**	83·5°—84·5°	83°[1]—91°[2]
(4)	Refractive index 40° C. (Z. B.)	**67**		
(6)	Solubility. Acetic acid	insoluble		
	Alcohol reagent	82°[3]		
(8)	Iodine value	**13**		
(10)	Saponification value	**80**	78—83	
(13)	Acetyl value	**55**		
(14)	Acid value	**2·5**	2—3	0·3[4]—7·0[5]
(15)	Unsaponifiable matter	**55**		

Remarks. **(4)** is taken at 80° C. and calculated to 40° C. **(10)** requires strong alcoholic potash. **(15)** consists of alcohols and hydrocarbons.

[1] Stürcke. [2] Schædler. [3] Fryer and Weston. [4] Berg. [5] Henriques.

Chemical Composition.

Not fully confirmed. *Stürcke* arrived at the following composition:

Esters:
- (1) Ceryl alcohol.
- (2) Myricyl alcohol.
- (3) A dihydric alcohol, $C_{25}H_{52}O_2$. M.P. 103·5° C.
- (4) Carnaübic acid[1].
- (5) An hydroxy acid, $C_{19}H_{38}\begin{cases}CH_2OH\\COOH\end{cases}$

and also a hydrocarbon, M.P. 59° C. In addition *Lewkowitsch* is of opinion that free Cerotic acid is present.

Adulteration.

Its high price offers a great inducement to adulteration. The *melting point* would be *reduced* by almost any adulterant; hence this is an important criterion of purity.

The commonest adulterants are Paraffin wax, Ceresin, and Stearic acid. The two former increase the percentage of unsaponifiable matter, and the latter the acid value.

Japan wax is determined by the percentage of glycerine yielded on saponification.

II. Technical.

Source.

From the leaves of a South American Palm (*Corypha cerifera*) found chiefly in Brazil.

Yield.

One hundredweight of wax is the produce of about 10,000 leaves.

Method.

The leaves are picked before fully extended, sun dried, the powdery wax removed from the dried leaves and melted over boiling water. The cake formed is broken and taken to the coast towns for shipment to Europe.

Refinement.

By remelting and straining. It is bleached by treatment with potassium dichromate or with Fuller's earth, and paraffin wax is often added, thus lowering the melting point.

Grades.

Yellow flor.	Fatty grey.
Chalky grey.	Bleached.

The chalky grey often contains much dirt and sandy matter.

[1] Cerotic acid? The chief ester present is apparently myricyl cerotate: myricyl alcohol exists also free.

Uses. Carnaüba wax, owing to its hardness and high melting point, takes a fine hard gloss on rubbing with a cloth. On this account it forms a valuable ingredient of shoe and furniture polishes. It is also a constituent of high melting-point candles, phonograph and gramophone records, and insulating materials.

Candelilla Wax.

§ 166. This wax is obtained from the stem of a leafless plant (the *Pedilanthus pavonis*), growing chiefly in Mexico. It is obtained by boiling the stems with water or by extraction of the wax with petroleum ether. It is brownish to greenish in colour with a characteristic odour on melting. It is not so hard as Carnaüba wax and does not take such a fine polish, but is employed as a substitute for it in the manufacture of boot polishes and for similar uses. The chemical composition is not fully known, but a hydrocarbon (Hentriacontane $C_{30}H_{62}$) has been isolated from it.

Melting point, 68°—70° C. Acid value, 10—20.

Specific gravity very variable, ·94—·99. Saponification value, 46—63.

Solubility. Acetic acid, True Valenta } insolubles.

Alcohol[1] reagent 63°[2].

§ 167. Other waxes in this group are Flax wax, Palm wax, Raphia wax, Gondang wax, Cane Sugar wax, for information of which the student is referred to more detailed works.

Animal waxes.

These are obtained from a great variety of sources and have little in common except their absence of glycerides. Typical Wax, *Beeswax*.

§ 168. WOOL WAX.

I. General and Analytical.

Character. Pale yellow translucent. Distinctive ODOUR. Unctuous CONSISTENCY.

Physical and Chemical Data.

	Test	Average figure	Usual variations	Extremes recorded
(1)	Specific gravity 15·5° C.	·945		
(2)	Melting point ° C.	35°	31°—41°	
(4)	Refractive index 40° C.	1·480	1·478—1·482	
(8)	Iodine value	25	17—26	
(10)	Saponification value	102		
(13)	Acetyl value	23		
(14)	Acid value (percentage)	50		
(15)	Unsaponifiable matter	43		

Chemical Composition.

Not fully known. Consists of a mixture of neutral esters and free

[1] Candelilla wax + 10 per cent. paraffin gave 72°,
" " + 33 " " " " 83°.

[2] Fryer and Weston.

alcohols, among which occur cholesterol and isocholesterol (page 62). Among the esters, LANOCERIN has been definitely proved.

Adulteration. Not usually adulterated. Glycerides and mineral waxes would be detected as under beeswax (q.v.).

II. Technical.

Source. Wool wax is the natural grease from the fleeces of sheep.

Preparation.

In the treatment of wool the grease is removed by treatment with weak soap or carbonate solutions, or by means of volatile solvents. In the first method the suds are run into a large tank and acidified. The grease rises to the surface and forms the "brown grease" or "recovered grease" of commerce. This is purified by processes (some of which are the subjects of patents) and forms, together with about 25 per cent. water, the "LANOLIN" of commerce. Another method is to distil the grease. The distillate is then pressed, forming "*distilled grease stearine.*"

Properties and Uses.

Although insoluble in water, it possesses the property of absorbing up to 80 per cent. of its weight of water. For this reason it forms a valuable medium for incorporating medicinal substances in solution to be used as ointment. It is also employed for superfatting toilet soaps.

§ 169. TYPICAL WAX. BEESWAX.

I. General and Analytical.

Character. Light yellow to dark greenish-brown (foreign). Sweetish characteristic ODOUR (often recalling honey). Fairly brittle, but plastic when warm.

Physical and Chemical Data.

	Test	Average figure	Usual variations	Extremes recorded
(1)	Specific gravity 15·5° C.	**·965**	·963—·966	·659[1]—·975[2]
(2)	Melting point ° C.	**63**	62—63·5	60·5[3]—76[3]
(4)	Refractive index 40° C.	**44**	43—45·5	
(6)	Solubility. Acetic acid	insoluble		
	Alcohol[4] reagent	76°[5]		
(8)	Iodine value	**9**	8—11	2·6[6]—49[3]
(10)	Saponification value	**95**	93—97	81·7[7]—150[3]
(13)	Acetyl value			
(14)	Acid value	**20**	18—21	4·4[3]—25·5
(15)	Unsaponifiable	**55·5**	55·3—55·6	

"Ratio-number" $\left(\frac{\text{sapon. value} - \text{acid value}}{\text{acid value}}\right) = \mathbf{3{\cdot}6}$ to $\mathbf{3{\cdot}8}$.

(4) Is estimated at 80° and calculated to 40° C.

[1] Deitze. [2] Hager. [3] Hooper. [4] Beeswax + 10 p.c. ceresin gave 82°.
[5] Fryer and Weston. [6] Buisine. [7] Buchner.

Note. Beeswax is imported from a large number of foreign countries. The above figures refer mainly, except in the "*extremes*," to the Home product. Foreign waxes give greater normal variations in the figures stated.

Chemical Composition.

Beeswax consists principally of MYRICIN (myricyl palmitate) and CEROTIC ACID. These are in the proportion of 6 : 1. In addition it contains small quantities of free alcohols—myricyl and ceryl—and of free melissic acid. About 12–14 per cent. of hydrocarbons are also present.

Adulteration.

Beeswax is very liable to adulteration ; besides the addition of other material after its production, it is now a common practice to make artificial honeycombs of paraffin wax or stearic acid and to place in the hive. The resulting wax will thus be largely adulterated.

PRELIMINARY TEST.

Dissolve a little in chloroform, allow to crystallise and examine crystals under microscope. Pure wax gives dumb-bell shaped tufts of curved needles. With 10 per cent. of paraffin wax or of tallow or stearine, the shape of the crystals is altered.

A. The *refractive index* is a useful sorting test. Taken at 84° and reduced to 40°,

Stearic	29—33
Japan wax	47—49
Carnaüba wax	65—69
Beeswax	42—45.

B. The *specific gravity* and *melting-point* are important indications of purity.

C. The *ratio number* is a good guide.

Carnäuba wax	39
Japan wax	11
Insect "	29
Spermaceti	Extremely high
Myrtle wax	68
Tallow	48
Stearic acid	0
Rosin	·18—·19
Paraffin wax	0
Ceresin	0

D. *Glycerides* determined by percentage of glycerol yielded on saponification.

E. *Stearic acid.*

On boiling with 80 per cent. alcohol and cooling (1 gramme—10 c.c.), filtering and mixing with water, stearic acid, if present, separates as flocks of crystals. If absent, only slight opalescence occurs.

F. *Ceresin and Paraffin.*

If present, and in absence of stearic acid, lowered acid and saponification values will be obtained. An important test is that of WEINWURM[1].

> Saponify 5 grammes of the wax with 25 c.c. $\frac{N}{2}$ alcoholic potash;
> evaporate off alcohol;
> add 20 c.c. glycerol;
> heat till solution occurs:
> run into 100 c.c. boiling water;

Pure beeswax gives a transparent and clear solution.
With 5 per cent. and upwards of hydrocarbons present, there is cloudiness.

G. *Rosin.*

Liebermann-Storch test. The percentage is determined by the Twitchell estimation (preferably after extraction of the Rosin and fatty acids—*see practical part*).

II. Technical.

Source.

Secreted by the *bee* in digestion and used for building honeycomb.

Method of Obtaining Wax.

The old method of obtaining the wax was by melting the combs in hot water. The wax was then strained to free it from suspended impurities.

The melted and strained wax is now usually expressed, and the press residue boiled up and repressed. A second residue, containing 10 to 15 per cent. of wax, is extracted by means of volatile solvents. This gives "extracted beeswax."

Refinement.

Beeswax is refined and bleached in the following manner:

(1) By continued melting in water.

(2) By sun bleaching. The wax is made into strips or ribbons after the addition of 5 per cent. of tallow and exposed to sunlight. Turpentine is sometimes used in place of tallow and probably acts as an oxygen-carrier.

(3) Treatment with ozone, potassium permanganate, potassium dichromate or hydrogen peroxide.

(4) A partial chemical bleach as in (3) followed by exposure to sunlight.

Bleached beeswax ("white wax") is heavier, more brittle, and has a smoother fracture than unbleached wax.

Uses.

Beeswax is mainly used in candle manufacture and in the preparation of wax polishes.

[1] *Chem. Zeit.* 1897, 519.

§ 170. Spermaceti.

Spermaceti is obtained as a solid precipitate from the head oil of the Sperm and Bottlenose Whales (yielding about 11 per cent. of wax) and also from Dolphin and Shark oils.

It occurs in glistening white masses, with crystalline structure. It is very brittle and can be readily powdered, and is, when pure, free from taste and odour.

It is refined by remelting, crystallising, and pressing and finally by treatment with weak caustic soda. Chemically it consists almost entirely of *cetin*, or cetyl palmitate. It is used for sperm candles, and for toilet purposes.

(1)	Specific gravity 15·5° C.		·960[1]
(2)	Melting point		49° C.[2] (recrystallised)
(6)	Solubility.	Acetic (standard)	61°[3] (deposits crystals)
		Alcohol reagent	44°[3] (deposits crystals)
(8)	Iodine value		3·5—4·0[4]
(10)	Saponification value		122—130[4]
(14)	Acid value		Not usually over 0·5 per cent.[4]

Adulteration is not usually practised, as the transparency of the wax is thereby impaired.

§ 171. Insect Wax.

Insect, or Chinese Wax is secreted by the "Coccus ceriferus," an insect of Western China. The wax is obtained by placing the larvae of the insect on certain selected trees up which the insect creeps, and on the branches and twigs of which the wax is deposited.

As much of the wax as possible is removed by hand and the twigs are then boiled with water, the wax, which floats to the surface, being skimmed off.

Insect wax resembles Spermaceti but has a more fibrous structure and is more opaque. Chemically it consists mainly of *Ceryl cerotate*.

It is not largely exported, and is employed in China and Japan for candle-making and in the manufacture of polishes.

(1)	Specific gravity 15·5° C.		0·970[5]
(2)	Melting point ° C.		81[5]
(6)	Solubility.	Acetic acid	insoluble
		Alcohol reagent	insoluble
(8)	Iodine value		1·5[4]
(10)	Saponification value		92[4]

[1] Dieterich. [2] Fendler. [3] Fryer and Weston.
[4] Lewkowitsch. [5] Allen.

BITUMEN WAXES

II. SOLID AND SAPONIFIABLE

(*b*) NON-ALCOHOLIC

CHAPTER XV

CLASS X. BITUMEN WAXES

§ **172.** The BITUMEN WAXES form a link between the vegetable waxes and the mineral waxes. The former contain esters, and usually either free acids, or alcohols, or both: the latter consist of hydrocarbons only. Bitumen waxes contain apparently esters[1] and free fatty acids, but no free alcohols. In this respect they resemble lignite and peat, the parent substances, which themselves are bodies intermediate between vegetable and mineral in character.

§ **173.** The Bitumen waxes are obtained from Thuringian lignite or *Brown coal*, Peat and similar substances by extraction with petroleum ether. That obtained from Lignite (Brown coal bitumen wax) is termed crude *Montan Wax*. It is brownish-black in colour, hard and brittle, breaking with conchoidal fracture. It has a lustrous appearance, and bituminous odour on friction or warming.

On solution of the wax in hot petroleum ether or petroleum spirit, the greater part of the wax is deposited on cooling, but about 25 per cent. remains dissolved and has a resinous character. It is almost insoluble in turpentine and most other solvents in the cold.

Refinement.

There are three patented processes for the refinement of this substance.

(1) **By distillation** with superheated steam in vacuo. This yields a light brownish product, which on pressing, gives a light brownish yellow wax, and on further pressing and treatment with animal char yields a white wax.

It has a fibrous structure, is fairly brittle, and consists chemically of MONTANIC ACID and an unsaturated hydrocarbon. The authors have

[1] On distillation, hydrocarbons are formed.

obtained the following figures for genuine waxes refined by this method:

Appearance—light brown to white.	
Melting point	72°—77° C.
Solubility acetic (standardised)	sol. at M. P.
Acid value	73—85.
Unsaponifiable matter	45—30.

The montanic acid, isolated from the wax, has the composition $C_{27}H_{55}COOH$, and belongs to the saturated series of fatty acids (see page 40).

The still residues are used for insulating purposes and in the manufacture of cylinders for phonographs.

(2) **Refinement by treatment with Sulphuric acid**[1].

The crude wax is melted with paraffin wax in the proportion of 30 parts of the former to 70 of the latter. The mixture is then treated with 10 to 20 per cent. of sulphuric acid of 66° Bé. (or stronger) at 160° C. to 200° C. with constant stirring until evolution of volatile matter ceases. It is then mixed with animal charcoal, filtered, and the wax employed without further treatment, or hot pressed to remove the paraffin wax.

Refined by this method the wax contains very much less free acid than that prepared by distillation (acid value 15—20), has a melting point of 77°—84° and resembles Ceresin wax closely in appearance.

(3) **Refinement by means of Nitric acid**[2].

This patented process consists in treatment of the heated crude wax with nitric acid (sp. gr. 1·2—1·4) for a definite time. The acid is removed by washing out with water. The wax is separated from the resinous product of the acid by filtration.

The characteristics of this wax are similar to the wax prepared by treatment with sulphuric acid, but it is purer and free from paraffin.

Ryan and *Dillon* examined a commercial sample of "montanin" wax. This consisted of Montan wax partially neutralised with soda (about ⅓ of the total acid so neutralised) the effect of which was to raise the melting point to 97° C.

Other bitumens yield similar products on like treatment.

§ **174. Peat wax** was obtained by Zaloziecki and Hausamann[3] by the extraction of peat with alcohol.

The Bitumen waxes, both crude and refined, are used in the manufacture of polishes, phonograph and gramophone records and for insulating purposes.

[1] German patents 200,050; 202,909; 254,701; 216,281; 260,697. English patent 1910. 22,500.

[2] German patents 207,488; 237 012; 247,357.

[3] French patent 338,736.

UNSAPONIFIABLE; LIQUID.

§ **175.** This group comprises the "Mineral oils," a loose term for the products of Petroleum, Shale and Coal, and for the oils obtained by the **destructive** distillation of lignite, brown coal, and other partially mineralised plant substances. It also includes the "Essential Oils," but, as stated in an early chapter, these lie outside the scope of this volume. [See however the chapter on Oleo-resins.]

MINERAL OILS

UNSAPONIFIABLE

LIQUIDS

CHAPTER XVI

CLASS XI. MINERAL OILS

§ 176. The Mineral oils, as their name implies, are those oils obtained either from the earth direct, or derived from earth products. These are detailed below. Petroleum, the most important, belongs to the first group; the remainder are derived from mineral products.

(1) Petroleum—a natural earth product.

(2) Shale tar oil—from destructive distillation of shale.

(3) Coal tar oil—from destructive distillation of coal.

(4) Lignite tar oil—from destructive distillation of lignite and allied substances.

Petroleum or **Rock oil** is a liquid occurring in the earth's crust in various widely distributed areas.

§ 177. Its *origin* has long been the subject of vexed discussion. Many chemists have been of the opinion that it has had an *inorganic* origin. Thus **Berthelot** in 1866 showed that the metallic carbides were acted upon by water giving rise to *hydrocarbons*. **Byasson** in 1871 ascribed its origin to the action of carbon dioxide and steam on iron or sulphide of iron at a white heat. Later the Russian chemist **Mendeléef** pointed out the probability of the existence of large quantities of iron in the earth's interior, and agreed with Berthelot in his theory of its formation by the action of water on carbides. In the year 1889 **Meunier** proved the presence of *ozokerite* in a meteorite, and the following year **Sokoloff** drew the conclusion that hydrocarbons are primitive earth substances which have been distilled by the interior heat of the earth and been condensed in the porous strata of the crust. Another theory is that of **Ross** (1891) who considers that the explanation of the origin of petroleum may be found in the action of volcanic gases (sulphur dioxide, sulphuretted hydrogen) on limestone (carbonate of calcium) resulting in the formation of hydrocarbons, with separation of sulphur and calcium sulphate.

The inorganic view of its origin has not been supported by geologists. They are on the contrary unanimous in ascribing an *organic* origin to it, and are of opinion that in most cases petroleum has been *formed in the strata in which it now occurs.*

Thus **Stery Hunt** pointed out that porous beds containing petroleum are often found lying between similar porous beds which are petroleum-free, and concludes that the petroleum must thus have been formed *in situ.* **Newberry** regarded petroleum as probably formed by the slow distillation of shales by the heat of the interior, but **Orton** points out that there is no evidence that even a temperature of 212 degrees Fahrenheit obtained at the depth of the lowest known shales.

In 1895 **Zaloziecki** showed that *adipocere*, the fat from corpses, contained fatty acids which on subjecting to heat and pressure yielded hydrocarbons.

Engler and **Höfer** both argue an *animal* origin for petroleum. Engler distilled Menhaden oil at a temperature of 300°—400°, and a pressure of several atmospheres, and obtained 60 per cent. of distillate containing naphthenes and olefines and from which he isolated a fraction which was indistinguishable from ordinary commercial kerosene. Höfer found little or no evidence of plant remains in petroleum beds, but obtained it in a coral reef. **Krämer** and **Spilker** (1900—1902) ascribed its formation to the action of heat on the waxy matter of diatoms, and **Lesquereux** to the decomposition of seaweed. The true view appears to be that different formations have had various origins, but that the Engler-Höfer theory of animal origin is probable in most cases.

The **distribution** of petroleum is world-wide, but the deposits differ in character and extent. The principal deposits occur along well-defined lines, often associated with the principal mountain ranges.

During the upheavals of the earth in past ages minor folds of the beds forming the earth's crust have been produced, which have arrested and collected oil volatilised from below—this in districts otherwise barren of oil.

The chief areas worked at the present day are situated in:

1. **America.** The principal fields are in Pennsylvania, California, Mexico, Texas, Canada, New Brunswick and Trinidad.
2. **Europe.** The most important fields are found in Russia, Galicia, Roumania, Hungary and Germany.
3. **Asia.** In the neighbourhood of the Caspian Sea, Sumatra, Borneo, Java, Burmah, Japan and Persia.

The following are approximately the proportions of the total petroleum produced in recent times by the various countries of the world:

America (States)	55
Russia	34
Sumatra, Java, Borneo	3·75
Galicia	2·75
Roumania	1·75
Burmah	1·5
India	·03
Canada	·25
Japan	·65
Germany	·30
Other countries	·02.

Thus Russia and the States together account for 89 per cent. of the world's output.

§ **178.** The **character** of crude petroleum varies greatly according to the locality.

A large amount of "natural gas" (i.e. low boiling point hydrocarbons) is invariably produced along with the liquid. A very large proportion of this is wasted as there are no means for its collection and transport.

Petroleum may be pale yellow in colour and of a limpid consistency, or brown, reddish brown to black, with green fluorescence, viscid to semi-solid in character. Between these extremes all gradations occur. The specific gravity varies from about 0·8 (·771 is known) to 1·06 (Mexican). The Flash Point from 0° to 280° (Abel Test). The odour is usually not unpleasant, but the Roumanian oils are distinctly agreeable, while those of Ohio, Algeria, Canada and Texas are offensive in character on account of the sulphur compounds they contain.

For percentage of solid hydrocarbons contained in crude petroleum of different origin see chapter XVII, § 187.

The chemical composition of the crude oils also varies considerably, and is more or less an indication of its geographical source.

§ **179.** The chief constituents of the more important petroleums are as follows:

A. AMERICAN.

The specific gravity of the crude oil varies from ·77 to ·82.

1. **Pennsylvanian.** The chief characteristic types of compounds found in these oils are

(i) Paraffin Hydrocarbons; these form the greater part.
(ii) Olefine Hydrocarbons
(iii) Benzene Hydrocarbons
(iv) Saturated cyclic hydrocarbons
} these all occur to a small extent.

2. **Californian.** The specific gravity of the crude oils varies from ·87—·88.

These oils differ considerably from the Pennsylvanian inasmuch as they nearly all contain nitrogen and more or less sulphur compounds.

The chief types of compound found in the oils are

(i) Paraffins; these occur to only a small extent and consist of those members boiling below 70° C.

(ii) Olefines; these hydrocarbons form the major portion of the hydrocarbons of the crude oil.

(iii) Benzenes; benzene and several of its homologues are always present.

(iv) Naphthalene; this hydrocarbon has been isolated from crude Californian petroleum.

(v) Asphaltic compounds; a large proportion of these bodies are present; they consist chiefly of carbon and hydrogen, but always contain sulphur. Their exact composition and constitution are not known.

(vi) NITROGEN COMPOUNDS; these are always present in the crude oils and are probably derivatives of pyridine C_5H_5N and quinoline C_9H_7N.

(vii) SATURATED CYCLIC HYDROCARBONS; small quantities of the series $C_n H_{2n-2}$ and $C_n H_{2n-4}$ have been obtained from the higher boiling fractions.

3. **CANADIAN.** The specific gravity of the crude oils varies from ·85—·88. These crude oils are characterized by their high content of sulphur.

The chief types of compounds occurring in the oils are

(i) PARAFFINS; these occur to a less extent than in Pennsylvania.

(ii) OLEFINES; to a small extent.

(iii) BENZENES; these hydrocarbons occur to a greater extent than in Pennsylvania.

(iv) ASPHALTIC COMPOUNDS; the amounts of these vary but they are the chief constituents.

(v) THIOPHENES; these are characteristic of Canadian crude oil and are cyclic compounds containing one atom of sulphur per molecule (see § 36).

4. **TEXAS.** The crude oils of this region appear to be intermediate between the paraffin-base bearing oils of Pennsylvania and the asphaltic-base bearing oils of California. They are also rich in sulphur, which is sometimes found in the free state.

B. EUROPE.

1. **CAUCASIAN.** The specific gravity of the crude oils varies from ·84 to ·88. These oils differ considerably in composition from those of American origin. The chief types of compounds occurring are

(i) NAPHTHENES; these form the major portion of all the crude oils (see § 31).

(ii) PARAFFINS; these occur to a small extent, and then only those of low boiling point; they contain less than one per cent. of those members which form paraffin wax.

(iii) ACIDS; these compounds form a very small percentage but are characteristic of the crude oils; they are probably acid derivatives of the naphthenes, and hence have been termed Naphthenic Acids.

Octonaphthenic Acid is

$$\begin{array}{l} CH_2-CH_2-CH-CH_3 \\ | \qquad\qquad\quad | \\ CH_2-CH_2-CH-CO-OH \end{array}$$

(iv) BENZENES; the lower members occur to a small extent.

(v) ACETYLENES; small quantities have been found.

2. **GALICIAN.** These oils are, generally speaking, intermediate between those of Pennsylvania and the Caucasus.

The chief types of compounds are

(i) PARAFFINS; these occur to a less extent than in Pennsylvanian oil, but contain more of the higher members producing a higher yield of paraffin wax.

(ii) NAPHTHENES. These occur to less extent than in Caucasian and more than in Pennsylvanian oils.

(iii) BENZENE HYDROCARBONS. To a slight extent.

3. **ROUMANIAN.** These oils are very similar to the Galician oils, but yield more of the lower boiling compounds.

C. ASIA. The composition of these oils has not been investigated so fully as those of America and Europe.

1. **BORNEO.** Two types of oil are obtained, viz.:
   a. A paraffin-base oil.
   b. An asphaltic-base oil.

2. **SUMATRA.** The oils of this district consist chiefly of paraffins with a high content of low boiling-point members.

3. **JAVA.** These oils in contradistinction to those of Sumatra, though consisting chiefly of paraffins, give a high proportion of the higher members capable of producing paraffin wax of a high melting point.

§ **180.** Petroleum, as such, is not an article of commerce, but is usually subjected to various processes whereby useful products are obtained. These products can be varied between certain limits by alteration in the conditions of treatment; during the last few years attempts have been made to increase the yield of the low-boiling point oils, e.g. petrol or gasoline. The following table gives the more important products commercially obtainable:

1. **AMERICAN PETROLEUM PRODUCTS.**

The products usually marketed by the American Companies may be conveniently divided into three main classes, but the individual products will vary with the locality from which the crude oil is obtained.

A. Light Oils. Those oils boiling up to 150° C.

	Name	S.G.	B.P.	Uses
1.	Cymogene		0° C.	For producing cold in refrigerators; not obtainable in this country.
2.	Rhigolene	·60	18·3° C.	Medicinal purposes; not obtainable in this country.
3.	Benzine	·638—·660	45°—60°	Solvent for rubber, fats, etc.; cleaning liquid.
4.	Petroleum Ether	·650—·666	70°—90°	Solvent for rubber, fats, asphalts; for making air-gas.
5.	Ligroin	·72—·74	120°—130°	
6.	Petrol	·75—·77	70°—120°	Internal combustion engines; solvent and cleanser. (Petrol and Gasoline)
7.	Gasoline		40°—70°	
8.	Benzoline		70°—95°	
9.	Naphtha		95°—120°.	

N.B. These names are very loosely used by different manufacturers and sometimes denote quite different products.

B. Burning Oils—Kerosene. Oils boiling from 150°—300°. These oils vary considerably with different firms. The different grades of oil are known by trade names, of which there are a great number. They are mostly graded according to **Colour** and **fire-test.**

The colours of the oils are classed as

1. Colourless termed Water-white.
2. Pale-yellow „ Standard-white.
3. Straw-colour „ Prime-white.

The following brands of burning oils, marketed by the Anglo-American Oil Co., are obtainable in this country:

White Rose, Royal Daylight, Crown Diamond, Finest American Lamp Oil, Petroleum Lamp Oil, oil for lamps, stoves, or oil engines, Mineral Colza Oil.

C. Lubricating Oils.

A great variety of oils are prepared for this purpose. They may be divided into two main classes, viz.:

1. Those that are to be mixed with animal or vegetable oils; these are generally of low specific gravity.

2. Those used alone. These cover a wide range of specific gravities and viscosities.

The following brands represent some of the lubricating oils marketed in this country by the Anglo-American Oil Co.:

Lubricating oil, Pale lubricating oil, American filtered lubricating oil, American red lubricating oil, Black machinery oil, Dark cylinder oil, Cylinder oil, Pale crude lubricating oil, etc.

In addition to these three classes of oils many firms produce two other important bodies, viz.:

(1) Vaseline; used medicinally, as rust protector on iron, lubricant, etc.

(2) Pitch; having a S.G. of about 1·20 and M.P. 84°—85° C. and used as a cement, for varnish making, asphalts and as a fuel.

2. **Russian Petroleum Products.**

The yield of **light oils** obtained from Russian petroleum is small; the chief object of the distiller is to obtain a large residuum of oil, suitable as a fuel, termed "Astatki" or "Mazut." The following products are those usually marketed.

A. Light Oils.

1. Light Benzine.
2. Heavy Benzine.

B. Burning Oils—Kerosene. S.G. about ·825.

1. Kerosene.
2. Solar Oil.

C. Lubricating Oils. S.G. ·86—·92.

The different grades are classified as Fine oil, Engine oil, Cylinder oil, Dark cylinder oil, Heavy cylinder oil, Spindle oil and Axle oil.

D. Astatki. S.G. ·88 to ·912; the usual yield of this product is from 50—60 % of the crude oil; its chief use is as a fuel in Russia.

3. **Galician Petroleum Products.**

The usual commercial products are obtained together with large quantities of paraffin wax. Ozokerite, a natural wax, is also extensively mined in Galicia (see § 189).

4. **Californian Petroleum Products.**

The Californian wells yield a large amount of gas which on subjection to condensation produces a good yield of petrol (gasoline).

Several of the usual commercial products are obtained, and in addition the residue, which is of an asphaltic nature, is used in the manufacture of artificial asphalt for paving purposes.

5. **Sumatra Petroleum Products.**

The chief product is petrol, though some of the usual commercial products are obtained.

6. **Borneo Petroleum Products.**

The crude oil, as previously stated, contains both paraffin bases and asphaltic bases.

The former yields the usual commercial products as well as a fair quantity of PARAFFIN WAX.

The latter gives good yields of FUEL OILS.

7. **Java Petroleum Products.**

These oils yield a large proportion of **solid paraffins** of high melting point, as well as some of the usual commercial products.

§ 181. SHALE TAR OILS.

These oils are obtained by the destructive distillation of bituminous shale and brown coal, the latter being chiefly worked in Germany, whilst the former is worked in Scotland, France and Australia; there are, however, vast deposits of suitable shale in various parts of the world, awaiting development.

The crude oil is a very complex body containing many types of compounds. The following table gives the usual constituents of the crude oil:

A. Constituents of the purified oil.

(i) **Paraffin hydrocarbons**	} These are the chief constituents.	
(ii) **Olefine** ,,		
(iii) **Naphthene** ,,	Small amount.	
(iv) **Benzene** ,,	Small amount.	

B. Constituents removed during purification and obtained as by-products:

1. BASES. (i) **Ammonia** recovered as Ammonium Sulphate.
   (ii) **Pyridine** and Quinoline derivatives.

2. HYDROXY-COMPOUNDS.
   (i) **Phenol.** C_6H_5OH.
   (ii) **Cresols.** $C_6H_4.CH_3.OH$.
3. HYDROCARBONS.
   (i) **Pyrene.**
   (ii) **Chrysene.**
4. TAR.
5. COKE.

§ 182. COMMERCIAL PRODUCTS OF SHALE.

Various products are obtained according to the mode of distillation and the method of purifying. The following table gives a list of typical products :

	S.G.	Uses
1. SHALE SPIRIT or NAPHTHA 3—5 °/₀ of crude oil.	·66—·75.	Fuel for internal combustion engines, solvent, making air-gas, etc.
2. BURNING OILS 20—30 °/₀ of crude oil.	·758—·830.	Lamps and engines.
3. INTERMEDIATE OILS 10—20 °/₀.	·840—·865.	For gas making and gas enriching, fuel for oil-engines, fuel for steam raising, grease-making, etc.
4. LUBRICATING OILS 20 °/₀.	·865 -·900.	For all lubricating purposes.
5. PARAFFIN-WAX 20 °/₀.		Candle-making, vaseline, matches, waterproofing, etc., etc.
6. COKE 3 °/₀.		Fuel.
7. TARS 15 °/₀.		After purifying from valuable bodies, as a fuel.
8. GASES.		Heating stills, illuminating, etc.

LIGNITE TAR OIL (Brown Coal Tar Oil).

This is obtained by the **destructive** distillation of Brown Coal or lignite. The Brown Coal is sometimes almost indistinguishable from coal, but is frequently of a woody structure. It occurs in strata of a more recent geologic age than coal (EOCENE, MIOCENE, OLIGOCENE, whereas coal occurs in the CARBONIFEROUS period). Saxony is the seat of the industry. If the Brown Coal is extracted with solvents, it yields **Montan Wax** (chapter XV). On destructive distillation, a tar is obtained of a buttery consistency, dark to light in colour, and either acid or alkaline in reaction. The yield of tar from the lignite varies from 2 to 10 per cent.

On fractionation the yields are as follows :

	Per cent.	B.P.	S.G.	Character
BENZENE	2—6	130°—170°	·71—·75	Chiefly **Benzene** and aromatic hydrocarbons.
PHOTOGEN	8—30	170°—220°	·75—·82	,,
SOLAR OIL	12—50	220°—290°	·82—·88	Unsaturated and saturated hydrocarbons.
LUBRICATING OILS	15—30	290°—320°	·88—·90	Mainly saturated (paraffins) hydrocarbons.
PARAFFIN WAX	1·5—12	above 320°	·89—·91	Paraffins.
Residue water and gases	22—50.			

§ 183. COAL TAR OILS.

The crude tar-oil obtained during the manufacture of coal-gas by the destructive distillation of coal is a very complex substance; the nature of the compounds forming the mixture, as well as the amount of each, varies with the kind of coal distilled and the temperatures at which the distillation has been carried out. The chief compounds occurring in crude tar-oil may be classified as follows :

I. HYDROCARBONS.

(i) **Paraffins.** A small amount of these are present, usually C_5H_{12} and C_6H_{14}.

(ii) **Benzenes.** These compounds form the major portion of the hydrocarbons present. The following members are usually present :

Benzene	C_6H_6	B.P. 81°
Toluene	C_7H_8	,, 110°
Xylenes (o, m, p,)	C_8H_{10}	,, o, 141°; m, 139°; p, 138°
Pseudocumene	C_8H_{10}	,, 169·5°
Mesitylene	C_9H_{12}	,, 163°
Durene	$C_{10}H_{14}$	,, 169°

(iii) **Naphthalenes.** These form the greater part of the solid hydrocarbons present. The following have been isolated :

Naphthalene	$C_{10}H_8$	M.P. 79°
α Methylnaphthalene	$C_{11}H_{10}$	B.P. 240°
β ,,	$C_{11}H_{10}$	M.P. 32·5°
Dimethylnaphthalene	$C_{12}H_{12}$	B.P. 262°

(iv) **Anthracene.** This is the most important of the remaining hydrocarbons :

Anthracene	$C_{14}H_{10}$	M.P. 213°

(v) Other **Hydrocarbons.** These occur to only a small extent:

Acenaphthene	$C_{12}H_{10}$	M.P. 99°
Diphenyl	$C_{12}H_{10}$	,, 71°
Fluorene	$C_{13}H_{10}$	,, 113°
Phenanthrene	$C_{14}H_{10}$	,, 100°
Fluoranthene	$C_{15}H_{10}$	,, 109°
Pyrene	$C_{16}H_{10}$	,, 149°
Chrysene	$C_{18}H_{12}$	,, 25°
Retene	$C_{18}H_{18}$	,, 98°
Picene	$C_{22}H_{14}$	,, 339°

2. OXYGEN COMPOUNDS.

These compounds are hydroxy-derivatives of the hydrocarbons and occur to a fairly large extent: they include the following:

(i) **Phenols.** These are the more important.

Phenol C_6H_5OH M.P. 42°

Cresols (o, m, p) $C_6H_4\langle^{OH}_{CH_3}$ o, M.P. 31°; m, B.P. 201°; p, M.P. 36°.

(ii) **Naphthols.**

α Napthol	$C_{10}H_7OH$	M.P. 84°
β ,,	$C_{10}H_7OH$	M.P. 123°

3. NITROGEN COMPOUNDS. These compounds are all complex derivatives of ammonia, which is always present to a small extent in tar-oil.

(i) **Pyridines.**

Pyridine	C_5H_5N	B.P. 114·8°
α Picoline (α methyl pyridine)	$C_5H_4(CH_3)N$	,, 130°
$\alpha\gamma$ Lutidine ($\alpha.\gamma$ dimethylpyridine)	$C_5H_3(CH_3)_2N$	,, 157°
(1, 3, 5)-Collidine (1, 3, 5-trimethylpyridine)	$C_5H_2(CH_3)_3N$	,, 172°

(ii) **Quinolines.**

Lepidine (γ methylquinoline)	$C_9H_6(CH_3)N$	B.P. 257°
Cryptidine (dimethylquinoline)	$C_9H_5(CH_3)_2N$	,, 274°

(iii) **Acridine.** $C_6H_4\langle^{N}_{CH}\rangle C_6H_4$ M.P. 110°

(iv) **Aniline.** $C_6H_5NH_2$ B.P. 182°

(v) **Other Compounds.**

Parvoline	$C_9H_{11}N$	B.P. 188°
Corindine	$C_{10}H_{15}N$	,, 211°
Rubidine	$C_{11}H_{17}N$	,, 223°

4. SULPHUR COMPOUNDS. These occur to a very small extent.

Thiophene	C_4H_4S	B.P. 84°
Thiotolene	C_5H_6S	,, 113°

§ **184.** Though tar-oil is a very complex liquid, the tar distiller does not attempt to isolate all the bodies it contains but confines his efforts to the isolation of a few very important ones.

The crude tar-oil is first distilled and the distillate collected in four fractions, viz.:

		B.P.	
1.	Light Oil	B.P.	up to 170°.
2.	Middle Oil	,,	170° ,, 230°.
3.	Heavy Oil	,,	230° ,, 270°.
4.	Green Oil	,,	over 270°.

These fractions are then worked up and from these the following commercial products are obtained:

1. **Benzols.**

(i) 50 per cent. Benzol; a mixture of Benzene, Toluene and Xylene of which 50% distils below 100° C.

(ii) 90 per cent. Benzol; a mixture of Benzene, Toluene and Xylene of which 90% distils below 100° C.

(iii) Benzene; pure benzene.

2. **Solvent Naphtha**; a mixture of benzene hydrocarbons.

3. **Naphthalene**; different grades are marketed varying in purity.

4. **Anthracene**; different grades are marketed varying in purity; usual brands are 40%, 50% to 90% anthracene.

5. **Phenol (Carbolic acid)**; grades vary from pure downwards.

6. **Cresols**; mixtures of three isomers varying in degree of purity.

7 **Creosote Oil**; a very complex mixture of hydroxy-compounds and hydrocarbons.

8. **Pyridine and Quinoline**; these are obtained in different degrees of purity.

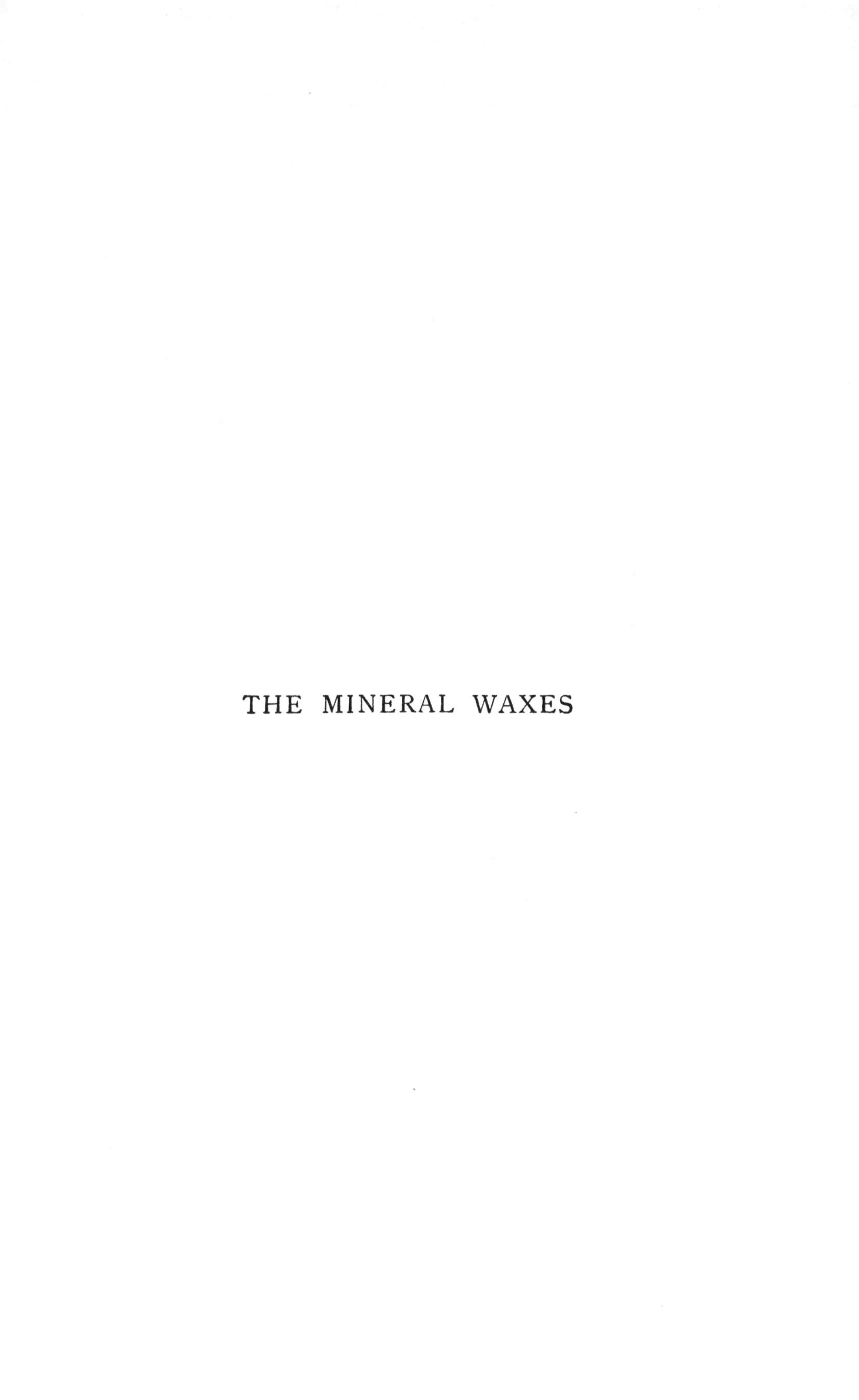

THE MINERAL WAXES

UNSAPONIFIABLE,

SOLID,

NON-ALCOHOLIC

This group comprises the **mineral waxes.**

They are all unsaponifiable on treatment with caustic alkalies, and they contain no alcohols, but consist entirely of **hydrocarbons.**

CHAPTER XVII

CLASS XII. THE MINERAL WAXES

§ 185. THE MINERAL WAXES, so called because of their occurrence in the earth's crust, may be divided into :

(1) Those obtained by **Distillation**—Paraffin wax from Petroleum and Shale.

(2) Those obtained direct by **Mining**—Ozokerite.

§ 186. PARAFFIN WAX is a mixture of saturated hydrocarbons of the C_nH_{2n+2} series. The following members have been shewn to be present in commercial samples :

		M.P.
Tricosane	$C_{23}H_{48}$	48°
Tetracosane	$C_{24}H_{50}$	50°—51°
Pentacosane	$C_{25}H_{52}$	53°—54°
Hexacosane	$C_{26}H_{54}$	55°—56°
Octocosane	$C_{28}H_{58}$	60°
Nonocosane	$C_{29}H_{60}$	62°—63°

The commercial samples of Paraffin Wax vary in their melting point, the softer varieties having lower melting points than the harder; the softer varieties not only contain lower members of the usual constituents but also more or less of the liquid members which have not been removed during the process of manufacture.

§ 187. The **yields** of Paraffin Wax from various sources are indicated under :

PETROLEUM.

Upper Burma	10—12 per cent.
Assam and Borneo	considerable proportion
Java, Timor, Algeria	rich in paraffin wax

PETROLEUM.	
U.S.A., Canada, Galicia, Roumania	moderate yield (about 2 per cent.)
Russia, Mexico	poor yield
SHALE.	10—15 per cent.
LIGNITE.	5—40 per cent.

Properties.

Commercial PARAFFIN WAX is a white to bluish white translucent, waxy material of lamino-crystalline structure. It is an extremely indifferent substance, being attacked only slowly by the strongest agents.

It is freely soluble in mineral oils, ether and benzene, but sparingly soluble in hot alcohol.

In turbidity tests, insol. boiling acetic acid and alcohol reagent.

The melting point varies from 35° C. to about 75° C., the higher melting points being obtained from the East Indian Wax.

It is oxidised by concentrated Nitric acid yielding *succinic* and *cerotic* acids. Chlorine attacks it slowly. Heated with Sulphur it carbonises with evolution of Sulphuretted Hydrogen.

VASELINE.

§ **188.** This is the registered name of a product introduced by the Cheeseborough Manufacturing Co. of U.S.A. It is obtainable from the American, Galician and Russian oils.

The usual method is to dissolve the residue in the stills, when this has reached a buttery consistency, in 6 parts of petroleum ether, add 1½ times the weight of animal char and treat for 1—2 hours. This is repeated until the liquid is bleached, when the solvent, after separation of the char is evaporated yielding an odourless, colourless or light yellow tasteless mass, with a blue fluorescence, of melting point 32° C.

It has no crystalline structure and gives no crystals from alcohol.

The yield is about 12 per cent. from Galician oils. On distillation, vaseline yields oils and *crystalline* paraffin wax.

It can be separated into a solid and a liquid by precipitation from ethereal solution with alcohol.

The hydrocarbons evidently differ in some way from those of the distilled wax.

Adulteration.

Many substitutes are marketed consisting of paraffin wax dissolved in oil. Often described as Vaseline, these are of course fraudulent imitations They may be recognised by crystallization from solvents.

OZOKERITE.

§ **189.** The **origin** of Ozokerite is still, like that of Petroleum, a matter of controversy. It is regarded by some as an intermediate product between natural fat and petroleum, while the commoner view is that it is an oxidation and condensation product of petroleum. As previously stated, it has been found in a meteorite (Meunier, 1889).

§ **190.** It **occurs** very widely distributed: but is mined principally in **Galicia**. It is also found in

Austria
Servia
Egypt
Tcheleken Islands
Orange Free State
Utah
etc.

Galician Ozokerite is principally mined in the neighbourhood of **Boryslaw**, the industry having commenced about the year 1860. The production has decreased in recent years Thus from 9000 tons in the year 1874 it rose in eleven years to 13,000 tons, but decreased during the following twenty years to 4,000 tons

§ **191. Properties.**

Ozokerite is very variable. It may be quite soft or as hard as gypsum. In colour it varies from a light yellow to a dark greenish brown.

Sp. gr. ·85 to ·95.

M.P. 58°—100° C.

Soluble in all oil solvents, and in boiling alcohol to extent of 3 per cent.

In turbidity tests, insol. acetic acid and alcohol reagent.

It is plastic when heated, and is electrified on rubbing.

The refined product is known as **Ceresin** (for description of Refinement, see § 232).

It differs from paraffin wax in being plastic and non-crystalline in character.

It consists of a mixture of hydrocarbons.

Melting point usually 65°—75° C.

§ **192. Uses.**

Ozokerite is employed as a substitute for beeswax, which it resembles in its plasticity. It is largely adulterated with paraffin wax, many so-called "ceresins" being in fact entirely paraffin wax.

The *melting point* is usually lowered by addition of paraffin, and this is compensated for by adding Rosin. The latter is however readily detected and can be estimated by ascertaining the *acid value*[1].

[1] It is safer to make also a Twitchell estimation of the Rosin, in case STEARIC acid is also present.

SECTION V

PRODUCTION AND REFINEMENT OF OILS, FATS AND WAXES

CHAPTER XVIII

FATTY OILS, FATS AND NATURAL WAXES

§ **193.** Oils and fats, as they occur in plant and animal tissues, are enclosed within the **cells** which form the basis of animal and plant structure. In order to free the oil it is generally necessary to rupture, in some manner, the enclosing cell membrane. This may be accomplished in the following ways:

1. By heat.
2. By grinding.
3. By pressure.

The modern method is to employ all three agencies to free the oil more or less completely from the tissue in which it is contained.

An entirely different method, and one which has only come into use in modern times, is to dissolve the oil by the action of **volatile solvents**, the latter being subsequently evaporated off.

The **waxes**, on the other hand, do not normally occur enclosed within cellular tissue. The vegetable waxes exist mainly in a free form on the surface of plants, while beeswax is deposited in the free condition as the honeycomb. "Sperm Oil" and "Spermaceti" are mostly contained in the cavities which occur in the head of the Sperm Whale.

§ **194.** As oils and fats are found and employed almost all over the inhabited globe it follows that the methods employed to prepare them will be very diverse. In those countries where little or no industrial progress has been made the primitive arrangements for obtaining the oils will still survive, while these will be superseded to a greater or less extent according to the influence which European civilisation, with the development of modern machinery and methods, has had upon the people concerned. In many cases, while oil is prepared by the natives of the country for their own use, a great amount of the oleaginous material is exported to Europe or America and there treated with modern appliances.

The history of the methods of oil production from the earliest prehistoric times is thus exemplified by the means employed in various countries in its preparation. Of these, the most primitive method of all consists in the use of natural agencies, as for instance, exposure to the heat of the sun, in the case of coconut, and the fermentation produced by the action of air and warmth in the case of Palm fruits. A similar case occurs in the Japanese fishing industry, where the fish are often allowed to rot in heaps until the oil exudes from the decomposing mass. A more enlightened method consists in the employment of **pressure** in some form, and the use of fire heat or boiling water as a means of freeing the oil from the material in which it is contained. Native methods, still in vogue in many parts of the world, represent the simplest forms of the modern processes of *rendering by heat*, and of *expression*. These will now be considered in detail.

RENDERING.

This method is applicable mainly to animal tissues from which the oil and fat is more readily yielded.

The old method of employing fire heat was open to many objections. Local overheating, resulting in burning and discolouration was almost unavoidable, and noxious volatile products were evolved which created a nuisance.

The use of boiling water, or steam in open pans, was subject, though in a lesser degree, to the same objections as to odour, and the employment of closed vessels, termed **digesters**, has now become very general.

A digester (of which an illustration is given on Plate II) consists of a vertical boiler, usually of iron or steel, in which the tissue to be rendered, after being chopped or sliced up, is placed on a false bottom, and steam at varying pressures is admitted. The oil is readily freed and floats on the surface of the condensed water, while the "greaves" or residual tissue, which may still contain fat, is usually treated again, using a small proportion of acid. The effect of the acid is to further set free the remaining fat by its coagulating ("cutting") effect upon the cell membranes.

Low temperatures are employed for fats intended for edible purposes in order to free the fat without any secondary action upon the animal tissue, which would result in the production of substances giving an unpleasant flavour to the fat.

In fats for industrial purposes such precautions are unnecessary and high pressure steam is used, resulting in a greater yield of fat.

A similar method is employed with **bones**, which are first broken in a special machine. The aqueous liquor resulting from the condensed steam contains in this case gluey constituents from the bones.

EXPRESSION.

The simplest form of expression, doubtless still employed by natives in many parts, was the use of heavy weights. The next step in advance was the invention of the **screw-press** and the **wedge-press**. The latter was in use until quite recent times in Europe and is still almost the standard method in the interior parts of China and Japan. One of the best forms was the **Stamper Press** or **Dutch Press**. This consisted of an iron box in which were placed bags of the seed separated by perforated plates. Pressure was applied by means of hardwood "stampers" which acted like pile drivers, forcing in wedges of hardwood at each stroke. Release of the pressure was obtained by forcing down an inverted or "key" wedge, when the exhausted cakes were easily removed.

All forms of presses have now been superseded in modern mill plant by the **Hydraulic Press**. The limit of hydraulic pressure is dependent only on the strength of the materials used in construction. The maximum pressure obtainable in practice is from 2 to 3 tons per square inch.

§ 195. MODERN OIL EXPRESSION.

The operations concerned in the expression of oil in modern oil-mills are as follows:

1. Receiving and storing the seed.
2. Cleansing the seed.
3. Removing shell (if necessary) or "decorticating."
4. Crushing the seed (forming "meats").
5. Heating ("cooking") the "meats."
6. Moulding the "meats."
7. Expressing the oil.
8. Clarifying or refining the oil.

Illustrations and descriptions of complete oil mills are given on Plates IV, V, and VI.

1. STORAGE OF THE SEED.

In order that no fermentative or "heating" action may take place in the seeds it is necessary that they be stored loose, and in perfectly dry condition. Light is excluded as much as possible, but there should be free access of dry air. If heating occurs, the flavour and quality of the oil will suffer (Plate III).

2. CLEANSING THE SEED.

This is accomplished by "screening," which consists in passing the seeds along an inclined cylindrical mesh fine enough to retain the seeds, but allowing dust and sand particles, etc., to fall through. This is followed by a similar treatment in a sieve of mesh coarse enough to just allow of the passage through the sieve of the seeds themselves, but retaining larger impurities, as pieces of wood, stones, or metal. In order to remove all traces of the latter, which would be very injurious to the rolls, magnets are sometimes used, over which the seeds pass, the magnet retaining any nails or other fragments of iron present (Plate IX).

3. DECORTICATION.

The shell or "hull" is removed in the case of Castor, Arachis American Cottonseed and some others. The shell is cut by means of knives contained on a revolving member, and adjusted so that only the shell is affected. The kernels then separate from the split hulls and the latter are removed by a current of air (see Plate XII)

4. CRUSHING.

This is accomplished by passing the seeds between chilled iron rolls. These are commonly arranged vertically and may be 4 to 5 in number. Guides are attached to each roll to conduct the seed between successive pairs of rolls. The seed is fed into hoppers above the apparatus (see Plates XIII, XVI).

5. HEATING (Cooking).

This operation is omitted in the case of "cold pressed" ("cold drawn") oils. With seeds of comparatively low value the object is to obtain a maximum yield of oil at one operation. For this purpose, heating is necessary in order to obtain the oil in a freer condition. The "kettle" in which the operation is conducted is provided with a steam jacket and a mechanical stirrer to mix the mass so as to ensure uniform heating. The precise temperature employed is a matter of experience, but 160°—170° F. is a usual one. The maximum temperature permissible is about 220° F. Steam is blown into the mass in order to moisten it to the requisite degree. "Over-cooking" renders the cake brittle, whilst under-cooking causes spueing of the "meats" in the press, and results in the formation of a "leathery" cake (see Plate XVII).

6. MOULDING.

The operation of moulding the cooked meal into cakes of the required size is necessary in the case of expression in presses of the "Anglo-American" type described in the following section. The method used is to press into a mould a measured amount of the cooked "meats" from the previous operation by means of a miniature hydraulic press termed the "cake former." A movable carriage receives the correct quantity of meal from the kettle and delivers it to the mould made of the requisite dimensions. A press cloth is folded over and the pressure is applied, forming a cake ready for placing in the press (see Plates XVII and XVIII).

7. PRESSING.

Hydraulic presses consist[1] in principle of two metal plates or tables, one of which is fixed and secured to the framework by means of strong pillars, and the other attached to the hydraulic ram and free to move

[1] Another type of press, which has not however come into favour, consists of a helical screw, operating in a conical cage in the manner of a milled soap **plodder**. It is continuous in action, but the output is comparatively small. Very high pressures are possible, thus enabling high yields of oil to be obtained from seed in the cold.

towards, and parallel with the first. The seed is placed in the interspace and the oil forced out by pressure between the plates.

Although both horizontal and vertical presses have been used, the former type is now practically obsolete. Presses are either **open**, in which the interspace between the plates is left accessible for placing in position the cakes or bags of meal, or **closed**, forming a compartment or cylinder in which the meal is loosely placed, and in which the ram operates as a piston.

Open Presses.

In the case of the **Marseilles** Press, the meal is filled into "scourtins," or plaited bags of hair, or leaves These are placed one over the other in the press and separated by means of loose corrugated iron plates. The weak point of this type is the difficulty of ensuring an even pressure, owing to the uneven building up of the material ; for this reason a very high pressure cannot be employed or there is rupturing of the "scourtins." This press is mostly confined, as its name signifies, to the south of France.

The **Anglo-American** Press overcomes these difficulties. In this type the interspace between the ram and the top plate is provided with a number (10 to 20) of iron plates, usually suspended evenly one from the other. A number of compartments of equal size is thus provided. Into these are placed the cakes of uniform size as they come from the "cake-former." On the application of pressure, the plates are forced together, thus squeezing the oil from the meal and causing its exudation through the cloths surrounding each cake. On release of the pressure the plates fall again into their correct position. The expressed cakes are then readily removed from the compartments (see Plates XIX, XX).

The plates are often embossed with the maker's name, or with some distinctive design, so that the pressure shall cause the cakes to receive the reversed impression (see Plate XXI).

Closed Presses.

These are known as **cage** or **clodding** presses. One type consists of a cylinder or cage built up of iron staves with interspaces for the efflux of the oil, and held together by means of steel hoops. In the other form the "cage" is a cylinder of steel perforated with holes at intervals. The disadvantage is its liability to become blocked by the meal choking up the holes. In both presses the meal is fed loose into the cage, and circular metal discs of accurate fit are placed over the meal at stated intervals. When the cage is full, the ram, acting as a piston in the cylindrical cage, causes exudation of the oil through the spaces provided (see Plates XXII, XXIII).

In both the closed and open types, the **pressure** is first applied gradually by means of low pressure pumps[1] or accumulators[2]. The

[1] See Plate XXIV. [2] See Plate XXV.

maximum low pressure is about 800 lbs. per square inch. Up to this point the greater part of the oil will be obtained. In order to complete the operation, another set of pumps[1] or accumulators[2] is brought into play giving a maximum pressure of about 2 tons per square inch in the case of the open presses, while in the closed presses, particularly those of the perforated type, about double this pressure can be obtained. These excessive pressures are not found necessary in practice, as the expressed cakes are required to contain about 10 per cent. of oil for feeding purposes. As the edges of the cakes are rich in oil, they are pared off in the cake-paring machine (Plate XXVII) and the parings ground and repressed.

Seeds containing a **high percentage of oil**, such as coconut, palm-kernel, castor, arachis, etc. are usually expressed in two or three stages. The first expression is often in the **cold**, yielding "cold drawn" oil of finest quality, this being followed by one or more expressions of the heated meal.

The first expression is usually in a press of the **cage** type, owing to the liability of spueing. Subsequent expressions are either in the cage or Anglo-American presses. In a modern type of the cage press the cage itself is in two sections, one of which is removable. A preliminary pressing is thus made, and the once expressed material, in its cage section, is transferred by a mechanical movement into another compartment of the press, where it receives a final expression (see Plate XXIII).

The oil flows from the press into tanks below the floor level.

For details of the mill operations the student should study the plates, with explanatory notes.

8. Clarifying and Refining the Oil.

The expressed oil is usually turbid owing to the presence of albuminous and mucilaginous matters. In the case of cottonseed oil, the oil has a dark ruby red colour. Oils are clarified by settlement (tankage) or by filter-pressing either alone or after treatment with Fuller's Earth.

In the case of cottonseed oil, the red colouration is removed by treatment with caustic soda solution (see page 130).

§ 196. EXTRACTION BY SOLUTION.

The highest yield of oil is obtained by this method, but the resulting "cake" is not fitted for cattle feeding, both on account of its low fat content, and also because traces of the solvent are very difficult to entirely remove. For this reason, extraction by volatile solvents is generally practised when the seed is naturally unsuitable for a feeding-stuff, or of very low value for this purpose (as in the case of castor), or when the seed is in a damaged or fermented condition.

The volatile solvents in use are:

(1) Petroleum ether.
(2) Carbon disulphide,
(3) Chloro-derivatives of hydrocarbons.

[1] See Plate XXIV. [2] See Plate XXV.

Other possible solvents, such as ether and benzene, are too costly, and in the case of ether too inflammable for industrial use.

The solvent is recovered by evaporation of the solution and condensing the vapours evolved. Not all the solvent however is recoverable. If this were possible, the question of the price of the solvent would be unimportant, as the expense would only be an initial one, and a given quantity would suffice for any number of extractions. Besides a certain amount which is always retained by the oil and the extracted seed, there is loss due to leakage of vapour through joints, imperfect condensation and other causes. It is of course the aim of manufacturers to reduce these losses to a minimum.

It therefore follows that the *cost* of the solvent is of the first importance, and it is for this reason that the more expensive chlorine substituted hydrocarbons, such as Carbon tetrachloride, and the Chlor-ethanes and Chlor-ethylenes have not come into favour commercially, although offering the great advantage of non-inflammability.

In considering the suitability of a solvent for industrial use, the following properties are important:

(1) SPECIFIC GRAVITY.

Since a given **volume** of solvent is necessary for the extraction of a charge of seed, it follows that a solvent of high density is proportionally dearer than one of a lower gravity. For instance, one ton of Petroleum ether of density 0·720 will go double as far in actual working as a ton of Carbon tetrachloride of density 1·632. For

1 ton Petroleum ether = 301 gallons,
1 ton Carbon tetrachloride = 137 „

It follows that in considering prices of solvents, the figure quoted should be multiplied by the specific gravity in order to get a true comparison of costs.

Solvents of a specific gravity higher than water such as carbon disulphide and carbon tetrachloride and its allies, are more easily protected from loss by evaporation since a film of water can be kept over the surface of the solvent. In the case of the first named, this is also of importance in lessening the chance of fire, as it is exceedingly inflammable.

THE SPECIFIC HEAT has a bearing on the cost question, since the greater the amount of heat required to raise the solvent to the boiling point, the more expensive it becomes in practice.

The LATENT HEAT OF EVAPORATION is similarly of importance because the working costs are raised and the solvent requires a comparatively large expenditure of fuel to distil it off during recovery.

The BOILING POINT. The same remarks apply here also, the lower the boiling point, the cheaper will be the cost of recovery. In the extraction of **bones** and substances containing water, a solvent of a boiling point higher than that of water is desirable, since in this case, by extracting at a temperature above 100° the water is removed, allowing the solvent to permeate the tissue more readily.

ACTION ON METALS. Since for extraction vessels and fittings of iron are preferable on account of cheapness, solvents which act on iron are in most cases undesirable. **Pure carbon disulphide** is without action on iron, but the commercial liquid has an action which has led to ignition of the solvent. **Carbon tetrachloride** attacks both iron and copper in the presence of moisture, necessitating the use of lead-lined vessels. The other **chlor-hydrocarbons** also attack iron, especially when water is present, producing hydrochloric acid which has a rapid corrosive action on the metal, and results also in the discoloration of the oil. Trichlorethylene has the least action, and has been employed in recent times in extractors of iron for off-coloured fats. **Petroleum ether,** properly refined, is without action on metals.

A table summarising these properties is given below:

Name	Formula	Sp. gr.	°C. Boiling point	Specific heat	Latent heat of evaporation	
Carbon disulphide	CS_2	1·29	46	0·16	79·9	Impure, attacks iron
Petroleum ether	C_nH_m	0·75	110—115	0·5	80	No action on metals
Carbon tetrachloride	CCl_4	1·63	76	0·21	46·4	Attacks iron and copper
[1] Trichlorethylene	C_2HCl_3	1·47	88	0·22		Attacks iron
[1] Perchlorethylene	C_2Cl_4	1·62	121	0·22		,, ,,
[1] Tetrachlorethane	$C_2H_2Cl_4$	1·60	147	0·27		,, ,,
[1] Pentachlorethane	C_2HCl_5	1·70	159	0·27		,, ,,

Owing mainly to its cheapness, **Carbon disulphide** is still largely used for extracting olive 'marc,' despite the fact that the oil thereby acquires a disagreeable flavour due to traces of the solvent. For most purposes however **Petroleum ether** is very generally employed. To be suitable for this purpose it needs careful fractionation, so as to distil over between a small range of temperature. Otherwise higher boiling fractions are likely to remain in solution in the extracted oil.

TYPES OF EXTRACTORS.

There are two types of extractors in use.

(1) Where the solvent is continuously distilled off, condensed, and returned to the extracting vessel, a relatively small quantity of solvent being thus employed many times over.

(2) Where a constant supply of solvent flows through a series of vessels containing the material to be extracted, and as the vessels become exhausted they are cut out of the series and recharged. The solvent issuing from the end extracting vessel contains a high percentage of oil. The solvent is distilled off and returned to the store tank.

Fig. 33, p. 249 shows a plant of the first type, and Plates XXVIII, XXIX and XXX photographs of the latter arrangement.

[1] These attack aluminium.

CHAPTER XIX

REFINEMENT OF OILS, FATS AND WAXES

§ **197.** In considering the question of the refinement of oils and fats it will be well, in the first place, to determine what are the impurities to be got rid of. Oils expressed in the cold from perfectly fresh seed need little or no refinement. Hot pressed oils acquire a higher flavour, darker colour and more pronounced odour. This must be due to changes produced by the heat on the oil whilst still contained in the plant cells. The substances which produce the colour and odour are albuminous in nature and are either in an extremely fine state of suspension, or are actually dissolved by the oil on heating. In the case of animal oils and fats, the conditions necessary for obtaining a high class product are the same, viz., freshness of material and a low temperature of rendering. Here again the impurities are most probably albuminous in nature, or are compounds resulting from the decomposition of albuminous matters.

§ **198.** In order to refine or effect a physical improvement of oils and fats, the following methods are employed :

(*a*) The albuminous bodies are destroyed by the action of *heat* at a temperature insufficient to injure the oil.

(*b*) They are precipitated by the action of *cold.*

(*c*) They are removed by the *absorptive action* of various powders in fine suspension.

(*d*) They are charred by the action of powerful *dehydrating agents.*

(*e*) They are carried down 'entangled' in *soap solutions* obtained by treatment of the oil with weak alkali.

(*f*) They are *oxidised* to innocuous substances by treatment with nascent oxygen.

(*g*) They are *reduced* to innocuous substances by treatment with hydrogen and a catalyst, or by other reducing agents.

(*h*) They are destroyed or rendered ineffective by the *action of light.*

The last three methods probably effect the improvement of the oil by altering the chemical composition of the impurities rather than by destroying or removing them. For this reason it will be more accurate to refer to the action as "bleaching" only, though a deodorising and sweetening action always to some degree accompanies the bleaching. Oils which are "bleached" in this manner are liable to "revert," the objectionable qualities returning by the changing back of the substances into their original form.

§ **199.** (*a*) ACTION OF HEAT.

Linseed oil is lightened and improved by this means. The temperature must not be raised above 220°—250° C., as otherwise thickening, due to polymerisation, results. The heating is best carried out by means of a steam coil with superheated steam, combined with a mechanical agitator. The use of air for agitation is inadmissible on account of the rapid oxidation which would take place at this temperature.

Other oils are improved by the treatment with steam. In this case, an "open" steam coil is used, and the steam blown through the oil by means of small holes in the pipe. Agitation is effected by the steam at the same time. Medicinal Castor Oil is refined by this method.

(*b*) ACTION OF COLD.

This method is only applicable to those oils whose content of solid glycerides is small. It depends upon the fact that the albuminous substances are less soluble in cold and are to some extent precipitated[1].

(*c*) ACTION OF ABSORPTIVE POWDERS.

The most commonly used are the various varieties of Fuller's earth and animal charcoal, or "decolorising carbon." China Clay and other dried clays have also an improving action, but to a less degree. The method is comparatively expensive, and is used principally for refining high class oils suitable for edible purposes. From 5 to 10 per cent. of the thoroughly dried earth or carbon is added to the oil rendered bright by heating with closed steam coils[2]. The vessel is shown in Plate XXXV. It consists of a mixer with preferably an archimedian screw agitator working in a sleeve. The mixer is jacketed for steam on the bottom in order to keep the charge at the desired temperature. This, with edible oils, should not exceed 80° C., otherwise an earthy flavour will be imparted to the oil. The agitation is continued from ½ to 1 hour, and the whole charge then pumped through a filter press[3]. The Fuller's earth retains a large amount of oil, even after prolonged "steaming out" of the filter press, and this may be recovered by treating with suitable solvents, although the oil so recovered is very dark in colour. Fuller's earth varies considerably in effectiveness, the grey varieties being usually preferable to the brown. Different oils are also affected to different degrees. Fuller's earth is probably to be preferred to "decolorising carbon," as it is more effective at the same cost for each.

The precise manner in which the earth or the char acts is not yet understood, but researches are being carried out with a view to a solution of the problem.

[1] See patents. German, 163,056. Niegemann.
English, 10,326, 1905. Buchanan.

[2] The maximum action is only obtained when all the materials are perfectly dry.

[3] See Plates XXXII, XXXIII, XXXIV.

Many waxes may also be refined by this method, and it is regularly in use for Carnaüba and Beeswax.

(*d*) CHARRING BY DEHYDRATING AGENTS.

The two agents normally employed are Sulphuric Acid and Zinc Chloride. If Sulphuric Acid is used over a certain concentration and above a defined temperature, the oil itself will be attacked. Under suitable conditions the acid has a selective action on the albuminous impurities, charring these only by removal of the elements of water. The char on subsiding carries down all impurities in suspension. Fig. 15 shows a suitable type of vessel for the treatment. It must be lead-lined and is preferably of conical shape at the bottom to facilitate the separation of the acid "foots" (sediment). It is provided with a "closed" steam coil and also an "open" coil for agitation with compressed air. The amount of acid used and its concentration must be ascertained by actual trials, as no definite figures can apply generally.

Thus with a crude whale oil containing a large amount of albuminous impurities and also entangled water, it may be found necessary to use the strongest commercial acid ("C.O.V."), as this is immediately diluted on addition to the oil. The temperature is never over 100° C., and usually lower temperatures suffice. If acid of insufficient strength be employed, there is great difficulty of separation, and "acid emulsions" occur which are exceedingly troublesome to deal with. If, on the other hand, the acid be too strong, the oil is badly discoloured; 1 to 3 per cent. of acid is the amount usually necessary.

The oil is first raised to the required temperature and the acid, diluted to suitable concentration, added in a slow stream during constant agitation of the oil by means of the air compressor; ½ to 1 hour usually suffices for the operation. The acid "foots" is run off next day and the clear oil washed if necessary with water or boiled up in another vessel with steam and allowed to settle. Many oils are refined by this method, which forms the preliminary treatment for the "Twitchell" deglycerination process. Any free fatty acids in the oil are rather increased than reduced in amount by this treatment, so that it is not by itself suitable for edible oils.

Zinc Chloride is an alternative refining agent. A concentrated solution must be used (S.G. 1·60 to 1·65) and the method is similar in all respects to the above. It is however much more expensive than the acid treatment and is on this account rarely employed.

(*e*) ACTION OF SOAP SOLUTIONS.

The soap is formed by the action of dilute alkali on the free fatty acids present. The impurities become partly dissolved and partly entangled in the soap liquid and are carried down with it. The method is more costly than the acid process, since a fair amount of oil is also carried down with the soap. This class of "foots" however can be readily treated for recovery of the oil (by addition of acid) or it may be completely saponified by boiling with more alkali, and the resulting liquor dried and sold as a low class soap. Since the free fatty acids are by this treatment removed, it is employed extensively in refining oils for edible purposes, such as Copra, Coconut oil and Cottonseed oil.

The operation can be conducted in an iron vessel, and agitation is carried out with compressed air, or in the vessel shown in Fig. 15. The amount of Caustic Soda added depends upon the percentage of free fatty acids present, but the strength will be from 50°—20° Twaddell. The alkali is well mixed, but not for too long, and allowed to settle. The possibility of emulsions forming is much greater than in the case of the acid treatment, but they can be readily separated by addition of strong brine.

A solution of soft soap acts much in the same way as the addition of alkali, but the emulsion danger is greater. Strong Carbonate solution may also be used if there is much free fatty acid present.

The "foots" may either be treated as above, or acidified, and the resulting mixture of oil and fatty acids distilled, producing a white fatty acid, suitable for candle material.

(*f*) TREATMENT BY OXIDATION.

This includes the following processes:

(i) Treatment with air.
(ii) " " ozone.
(iii) " " peroxides, persulphates, perborates, etc.
(iv) " " potassium or sodium dichromate.
(v) " " nascent chlorine. (Generated from bleaching powder, or by sodium dichromate and hydrochloric acid.)

Although this process is usually referred to as a "chlorine bleach," the action is in reality one of oxidation. Chlorine of itself is absorbed by a fat producing chlorine addition and substitution products. *Nascent* chlorine (chlorine newly liberated in the atomic condition) acts upon water producing *nascent* oxygen as shown in the following equation:

$$2\mathbf{Cl} + H_2O = 2HCl + \mathbf{O}.$$

As in the case of sulphuric acid (v.s.) it is found that the albuminous impurities and colouring matters in oils are more readily acted upon than the glycerides themselves. It is thus possible to bleach oils by oxidation without appreciably oxidising the glycerides.

(i) **Treatment with Air.**

This method is used for bleaching Palm Oil and for improving and clearing many oils and fats. The oil is placed in an iron vessel or lead-lined timber tank, and raised to a temperature of 100°-150° C. by means of a 'closed' steam coil. Air is then blown through the oil by means of an 'open' coil from an air compresser, or the arrangement shown in the diagram may be employed. This consists of an ordinary steam injector (*A*) affixed to a closed vessel, into the bottom of which the air pipe passes. The action of the injector sucks air through the coil and keeps a constant stream of air bubbles passing through the oil. (*B*) is the closed steam heating coil, and (*C*) the draw off cock. The treatment is continued only until the oil is bleached, to prevent oxidation of the glycerides.

APPARATUS FOR BLEACHING OF PALM OIL BY MEANS OF AIR

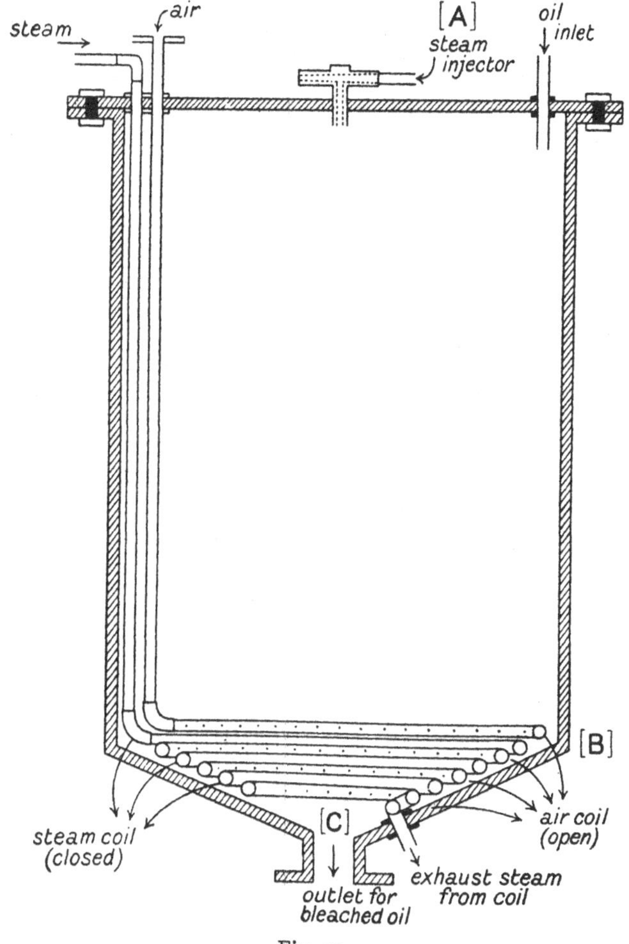

Fig. 15.

(ii) **Treatment with Ozone.**

This method has also proved successful for bleaching and deodorising oils. Ozonised air is produced by the silent discharge of electricity through air. The ozonised air is then passed through the oil in an apparatus similar to Fig. 15.

(iii) **Treatment with "per" Salts and Oxides.**

The "per" salts, such as Potassium and Ammonium persulphate[1], the perborates and percarbonates—all bodies which under the action of heat liberate nascent oxygen—are very powerful oxidising bleaches for oils, but are more commonly used for bleaching **soaps** in which they are soluble. They are relatively expensive compared with most other agents owing to the difficulty of manufacture[2].

The peroxides are also powerful bleaching agents but, with the exception of hydrogen peroxide, must be used with extreme circumspection owing to the violence of their reaction on contact with organic material[3].

The reactions are as follows:

$$2KSO_4 + H_2O = 2HKSO_4 + O.$$

$$\text{potassium persulphate} + \text{water} = \text{potassium hydrogen sulphate} + \textit{nascent}\ \text{oxygen.}$$

$$Na_2O_2 + H_2O = 2NaHO + O.$$

$$\text{sodium peroxide} + \text{water} = \text{sodium hydrate} + \textit{nascent}\ \text{oxygen.}$$

In addition to these, various organic peroxides have been employed under patent names for bleaching fats. Some of these dissolve in the oil and are thus superior to water soluble bleaches whose action depends upon the close physical contact of two immiscible liquids. Instances are benzoyl peroxide[4] and acetone peroxide.

(iv) **Bleaching by Sodium Dichromate.**

This method is largely used for bleaching tallow and palm oil. The treatment must be carefully carried out, and the temperature kept low, or there is a great tendency to discolour the fat (which is then termed "*foxy*" in works' parlance). A lead-lined tank is necessary, iron being unsuitable on account of the action of the reagents upon it.

The oil is first heated to about 50°—55° C. The dichromate ($1\frac{1}{4}$ to $1\frac{1}{2}$ per cent. of the oil should be sufficient) is dissolved in a minimum quantity of water, added to the fat and the whole well mixed by compressed air; hydrochloric acid ($2\frac{1}{2}$—$2\frac{3}{4}$ per cent. of the oil) is now added, and the whole agitated for a few minutes. The bleached oil is separated from the dark chrome liquor, and washed with water. Sulphuric acid may also be used but not so successfully. The chemical reactions are as follows:

1 "Palidol." German patents 200,684, 205,067.

2 The "per" salts are obtained by electrolysis of the corresponding normal hydrogen salts, the nascent oxygen liberated at the anode acting thus:

$$2KHSO_4 + O = H_2O + 2KSO_4.$$

$$\text{Hydrogen potassium sulphate} + \textit{nascent}\ \text{oxygen} = \text{water} + \text{persulphate of potash.}$$

3 Fatal accidents have attended their use.

4 German patent 214,937.

$$Na_2Cr_2O_7 + 14HCl = 2CrCl_3 + 2NaCl + 7H_2O + 6Cl.$$

		$2CrCl_3$	$2NaCl$	$7H_2O$	$6Cl$
		chromic chloride	sodium chloride	water	*nascent* chlorine

$$2Cl + H_2O = 2HCl + O.$$

$2Cl$	H_2O	$2HCl$	O
nascent chlorine	water	hydrochloric acid	*nascent* oxygen

$$Na_2Cr_2O_7 + 5H_2SO_4 = Cr_2(SO_4)_3 + 2KHSO_4 + 4H_2O + 3O.$$

In place of dichromate, manganese dioxide and potassium permanganate may be used. The reactions are then as follows :

$$MnO_2 + H_2SO_4 = MnSO_4 + H_2O + O.$$

MnO_2	H_2SO_4	$MnSO_4$	H_2O	O
manganese dioxide	sulphuric acid	manganese sulphate	water	*nascent* oxygen

$$K_2Mn_2O_8 + 4H_2SO_4 = 2MnSO_4 + 2KHSO_4 + 3H_2O + 5O.$$

$K_2Mn_2O_8$	$4H_2SO_4$	$2MnSO_4$	$2KHSO_4$	$3H_2O$	$5O$
potassium permanganate	sulphuric acid	manganese sulphate	potassium hydrogen sulphate	water	*nascent* oxygen

(v) **Nascent Chlorine.**

As in last section (iii) with dichromate and nydrochloric acid, or using bleaching powder.

With the latter, the same remarks apply as with the dichromate, the bleaching powder being first added in concentrated solution (0·1—0·2 per cent. of dry bleaching powder on the oil) and thoroughly mixed with the oil, and the hydrochloric acid added while the agitation is proceeding (0·3—0·6 per cent. on the oil). The optimum temperature is 50°—75° C. The charge is allowed to settle after about 15 minutes agitation, and the acid liquor run off from below. It is necessary to see that the bleaching powder is **fresh** and **dry**, otherwise it is largely ineffective. The reaction is as follows :

$$Ca(ClO)_2 + 4HCl = CaCl_2 + 2H_2O + 4Cl.$$

$Ca(ClO)_2$	$4HCl$	$CaCl_2$	$2H_2O$	$4Cl$
calcium hypochlorite (the active part of bleaching powder)	hydrochloric acid	calcium chloride	water	chlorine (*nascent*)

$$2Cl + H_2O = 2HCl + O.$$

$2Cl$	H_2O	$2HCl$	O
nascent chlorine	water	hydrochloric acid	*nascent* oxygen

(*g*) TREATMENT BY REDUCTION.

Treatment with sulphur dioxide, in a similar manner to the air treatment, has a bleaching action on most oils. The colouring matters, in this case, are partially reduced to colourless substances. More powerful action is obtained by the use of "hydrosulphites," particularly "formaldehyde hydrosulphites." Patents have also been taken out for the following compounds for use as oil and soap bleaches :

"Formaldehyde-sulphoxylate." German patent, 214,043.

"Sulphoxylates." English patent, 11,983 (1906) and several other foreign patents[1].

"Hydraldite[2]."

In the process of hardening of oil by hydrogenation with hydrogen and a catalyst (see chapter XX), the fat produced is largely bleached and deodorised. Unfortunately this action is incomplete before the glycerides themselves are reduced so that it is not a feasible method to employ for oils *as* oils.

(*h*) ACTION OF LIGHT.

The action of light on fats and waxes has already been discussed (§ 23). Direct sunlight is a powerful bleaching agent, while diffused daylight acts more slowly. Owing to the fact that the action is comparatively slow, it is only used in practice for the finer oils, such as medicinal castor oil, salad oils, oils for artists (poppy, walnut, linseed).

The oil is exposed under glass in shallow trays, or is poured into narrow bottles. Beeswax is also bleached in this manner (see page 187).

Patents have also been taken out for bleaching by exposure to ultra-violet rays, such as those evolved from the Cooper Hewitt[3] mercury vapour lamp, and from the "Uviol" lamp.

[1] French. 431,294 and 410,824.
German. 224,394.
English. 16,260 (1909), 12,157 (1911), 22,453 (1911), 3,433 (1912). 21,359 (1911).

[2] Sodium hydrosulphite + ammonia.

[3] German patents 195,663, 223,419.

CHAPTER XX

HARDENED FATS

§ **200.** It is not within the scope of this book to describe the many attempts made to obtain substances of high melting point from oils for the manufacture of candles. Some of these have been successful and are worked on the large scale, and some have failed after a practical trial. The "hardened fats" so produced were however generally unsuitable for other purposes than candle-making, and a detailed description of their manufacture belongs to that department of the fats industry.

In recent years, however, a "hardened fat" has been produced in large quantities commercially which resembles other fats in all essential respects. It has in fact been obtained by conversion of the unsaturated glycerides into saturated fats, by the direct addition of hydrogen. This was a direct result of the now classical researches of the chemists *Sabatier* and *Senderens*, who published their results in the year 1900. They found that, although hydrogen was not absorbed directly by unsaturated bodies of the fatty series, the action was rendered possible by the influence of **metallic catalysts**.

§ **201.** The phenomenon of **Catalysis** is very common amongst chemical reactions and has been recognised, under different names, from early times. Students of chemistry are familiar with the fact that oxygen is liberated from potassium chlorate by the action of heat more easily and freely in the presence of manganese dioxide than in its absence, and also that the manganese dioxide remains chemically unchanged after the heating. Again the union of sulphur dioxide gas with oxygen to produce sulphur trioxide takes place rapidly when the mixture is heated in the presence of finely-divided platinum which however undergoes no chemical change. In these reactions the manganese dioxide and the finely divided platinum are said to act catalytically and are termed catalysts.

Many hypotheses have been advanced explaining how the catalyst acts in these reactions but a discussion of these hypotheses would lead to no useful result[1].

[1] The reader is referred to Mellor's *Chemical Statics and Dynamics* (Longmans) where a very exhaustive treatment of the subject is given.

The following facts, concerning catalysis, may however be noted:

1. The catalyst has the same chemical composition at the end of the reaction as at the beginning though it may not be in the same physical condition.

2. A small quantity of the catalyst is able to assist in the chemical change of an indefinite amount of the reacting substance or substances.

3. A catalyst is incapable of starting a chemical reaction; the function of the catalyst is to change the rate of the chemical reaction[1].

DIAGRAM OF APPARATUS FOR THE HYDROGENATION OF AN UNSATURATED LIQUID CARBON COMPOUND

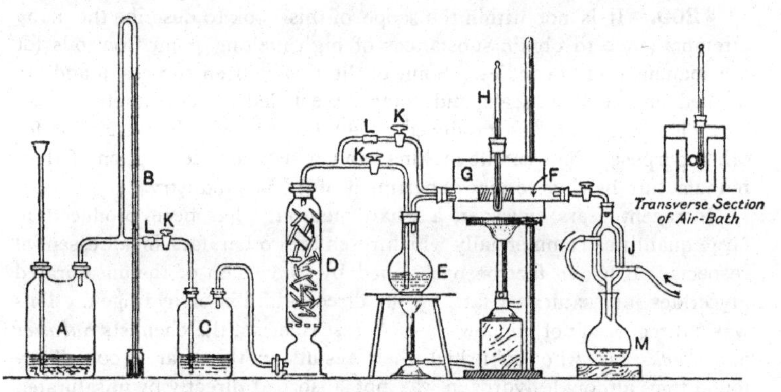

Fig. 16. A. Hydrogen generator—zinc and dilute sulphuric acid and small quantity of copper sulphate solution. B. Mercury safety valve. C. Saturated solution of potassium permanganate. D. Tower of solid caustic potash. E. Liquid to be hydrogenated. F. Hard glass tube containing reduced nickel. G. Hot air bath. H. Thermometer, the bulb of which touches tube F. J. Double surface condenser. K, K, K. Taps for controlling direction of current of hydrogen. L, L. Thick-walled rubber connectors. M. Receiver for hydrogenated compound.

4. Certain substances have the power of inhibiting the action of catalysts; the catalyst is commonly said to be "poisoned" by these bodies; thus certain sulphur and arsenic compounds cause platinum to lose its catalytic power in the union of sulphur dioxide with oxygen.

§ **202.** Sabatier and Senderens experimented with the **vapour** of the unsaturated body, and the absorption of hydrogen occurred when the vapour of the substance, together with hydrogen, was brought into contact with the catalyst.

[1] This statement is to be received with a certain amount of reservation, as several reactions are known in which no evidence has yet been adduced that the reaction takes place without the aid of the catalyst.

An apparatus suitable for the Sabatier and Senderens reaction is shown in Fig. 16. *A* is a hydrogen producer, *C*, *D* are wash bottles, *E* is a vapouriser containing the substance to be hydrogenised, *G* is the hydrogen chamber containing the nickel catalyst and heated by an air-bath, and *J* is a condenser for condensing the vapours of the hydrogenised material.

They were completely successful in reducing—i.e. saturating with hydrogen—many unsaturated bodies.

Oleic Acid, in the form of vapour, is converted thus :

$$\begin{array}{l} CH_3.(CH_2)_7.CH \\ \qquad \| \\ \qquad HC.(CH_2)_7.CO.OH \end{array} + 2H$$

oleic acid 2 atoms hydrogen

$$= \begin{array}{l} CH_3.(CH_2)_7.CH_2 \\ \qquad | \\ \qquad H_2C.(CH_2)_7.CO.OH \end{array}$$

$$\text{or } CH_3.(CH_2)_{16}.CO.OH.$$ stearic acid

To these experimenters therefore belongs the honour of solving what had long been a fascinating problem in the oil industries. The practical importance of the conversion of oleic into stearic acid depended upon the increased value of the latter as a soap and candle material. All attempts at directly uniting hydrogen and the higher unsaturated fatty acids had failed before these chemists obtained the result by catalysis[1].

A point of great importance is the relatively small amount of hydrogen required. In the equation above, 282 parts by weight of oleic acid are converted into stearic acid by 2 parts by weight of hydrogen. Thus 1 ton of oleic acid requires only about 16 lbs. of hydrogen or about 2,800 cubic feet of gas.

§ **203. Patents** were taken out based on the results obtained by the French chemists. Thus *Schwoerer* passed hydrogen and the vapours of oleic acid obtained by use of a jet of superheated steam, into a chamber in which was contained the catalyst embedded on a basis in a helical or screw form over which the vapours passed. Another method was patented by *Bedford* and by *Day* although the latter dealt with the saturation of hydrocarbon oils.

§ **204.** These methods all had reference to the **vapours** of fatty substances. They were therefore of little use in the case of glycerides, which cannot be vapourised without undergoing decomposition. In

[1] As early however as 1875 the reduction of oleic to stearic acid was accomplished by Goldschmidt by means of amorphous phosphorus and hydriodic acid at a temperature of 200°—210° C. An attempt to use the method on a practical scale failed owing to the loss of iodine involved. The use of chlorine was attended with great difficulty on account of its violent action on metals.

the year 1903 a patent was granted to **Normann** which describes the hydrogenation of oils in the *liquid* condition, by merely passing a current of hydrogen into the oil, immersed in an oil bath, and containing finely-divided metal as catalyst.

This was the patent which was destined to form the basis of the well-known legal action. It was held to be fundamental in character, and hence of immense value to the holders, as it precluded all other manufacturers from hydrogenating oils in the liquid condition. This view was not however upheld owing to the meagre information given in the patent; the work of Lewkowitsch, for instance, having shown that a successful result was very far from assured[1] working in the manner stated, and on March 18, 1913, after a four-weeks' action, and the hearing of a vast amount of evidence, a decision was given against the validity of the Normann patent.

§ **205.** The main essentials for success in the hydrogenation of oils in the liquid condition may be briefly stated as under:

(1) The correct control of the *temperature.*

(2) An adequate *pressure* of the hydrogen gas. The higher the pressure employed up to a point, the more rapid and certain is the hydrogenation.

(3) The fine state of *division* of the catalyst.

(4) The *purity* of the reagents.

(5) The exclusion of certain substances which destroy the action, and are said to "*poison*" the catalyst. The chief of these are·

Chlorine,
Sulphur (as sulphide),
Arsenic,
Phosphorus.

Of late years the most important factoı of alı nas proved to be the PHYSICAL CONDITION OF THE CATALYST ITSELF. If the catalyst is sufficiently active an almost instantaneous conversion of the oil is possible.

§ **206.** Although the action of the hydrogen is, and can only be, by **solution** in the oil, and the catalyst, once moistened by oil, cannot come into contact with the gas except in solution in the oil, yet many patents have been taken out describing devices to attempt to bring the catalyst, oil and hydrogen into further intimate contact by violent agitation or "atomisation."

All that is apparently needed is a gentle mixing of the contents, and even this may be unnecessary in view of the rapid diffusion which takes place.

A mass of patent literature exists in connection with the subject, but only two representative types of apparatus will be described here. For further information recourse may be had to the original patents in detail.

[1] Lewkowitsch failed to obtain any reduction prior to the publication of Normann's patent.

The **Testrup** patent (English 7726, 1910) describes the apparatus shown on Fig. 17. This consists of a series of closed vessels in which the oil and catalyst are sprayed in series into hydrogen gas, at a constantly decreasing pressure. Thus in the vessel (*A*) in which the oil is brought to the requisite temperature (160°—170° C. is mentioned) the pressure of the hydrogen is say 15 atmospheres. In the first spray chamber (*B*) hydrogen is admitted at a pressure of say 12 atmospheres. This causes the spraying of the liquid (which is at 15 atmospheres pressure) into this chamber. The oil is then conducted to the next of the

DIAGRAM OF TESTRUP APPARATUS FOR HYDROGENATION OF OILS

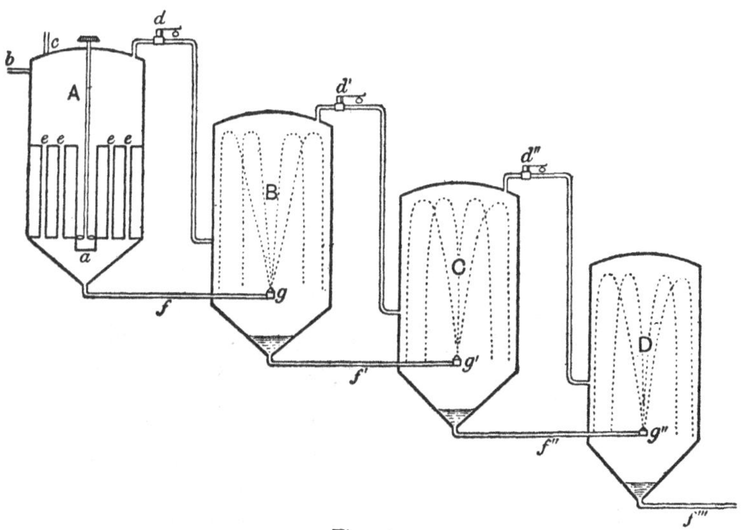

Fig. 17.

series (*C*) in which the pressure of the hydrogen is reduced to, say, 9 atmospheres, the difference in pressure causing the oil to again become projected in a spray. The oil and catalyst are in this manner sprayed ten or fifteen times into gas at varying pressures.

Two or three per cent. of catalyst (nickel or palladium) is used and agitation with the oil is effected by means of the stirrer at (*a*). The oil is admitted at (*b*) and the gas at (*c*), (*d*) being the reducing valve for regulating the pressure of the gas in the succeeding chamber, (*e*) (*e*) (*e*) (*e*) are heating tubes for superheated steam, (*f*) the pipe conveying the oil to the spray nozzle (*g*).

§ **207.** The **Kayser** process (U.S. patent 1,004,035, Sept. 26, 1911) consists in mechanically agitating the oil and catalyst in presence of hydrogen under pressure. The apparatus figured in the patent is shown in the adjoining illustration (Fig. 18).

It consists of a jacketed drum, placed horizontally through which, closed by stuffing boxes at either end, works a shaft bearing paddles covered with gauze.

A is a compression pump drawing hydrogen from pipe *B* through *C* to the treatment vessel. The oil and catalyst are admitted until the

DIAGRAM OF KAYSER APPARATUS FOR HYDROGENATION

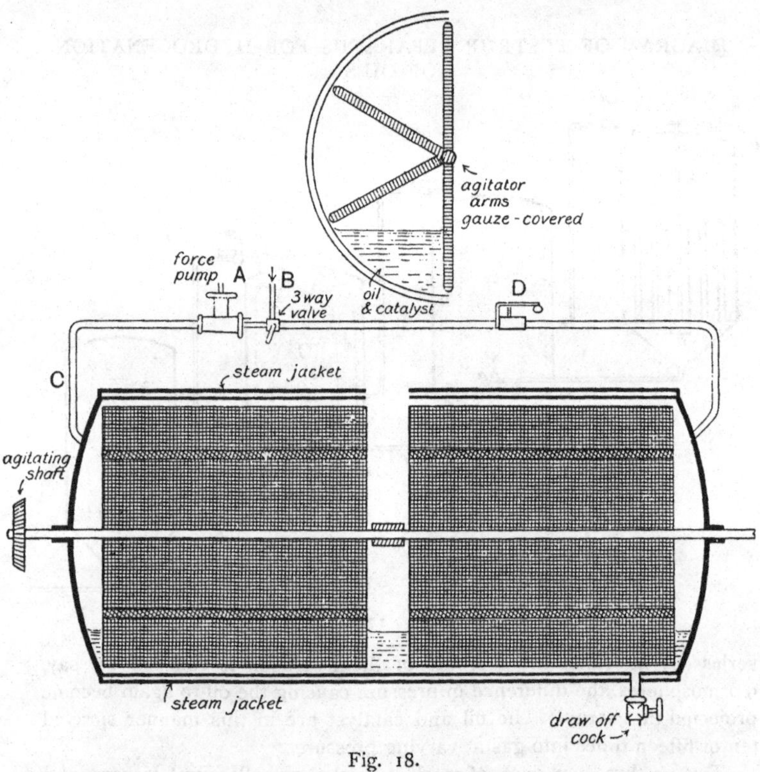

Fig. 18.

vessel is $\frac{1}{4}$ to $\frac{1}{3}$ full. Unabsorbed hydrogen is blown off at *D*. *E* is the draw off cock. The temperature mentioned for treatment is 150°–160° C. In a later patent the use of a carrier for the catalyst is claimed, this being an absorbent powder such as *kieselguhr* on to which the metal is precipitated.

§ **208.** As before stated, the most important point in the process is the **preparation of the metallic catalyst**. In earlier experiments,

Palladium was the metal used, and although very small quantities only are necessary[1], its relatively high cost compared with Nickel and the unavoidable loss entailed have practically displaced its use in favour of the latter.

Besides Nickel (and Palladium) the metals Platinum, Copper and Iron have been employed to a small extent.

The *activity* of the catalyst depends upon

(1) **Its fine state of division.**

It has been shown that it is necessary for the catalyst to be in the condition known as *pyrophoric*. In this condition the metal is in such a fine state of division that it spontaneously inflames on contact with air or oxygen. This physical condition is obtained by reduction of the oxide at *low temperatures* in a current of hydrogen. The oxide begins to reduce at 200° C., but reduction is incomplete below 270° C. and a good working temperature is 300°–320° C.

(2) **The purity of the catalyst.**

Nickel catalyst was formerly prepared from the sulphate, but the difficulty of entirely removing sulphur compounds, which affected the activity of the catalyst led to its abandonment in favour of the **nitrate**. The hydroxide is obtained by precipitation with an alkali, and after careful washing the precipitate is dried and reduced as stated above.

(3) **Protection from Oxidation.**

After the reduction of the nickel oxide it must not be allowed to come into contact with the air, and is best kept in an atmosphere of hydrogen until time of use. *Erdmann* and *Bedford* used the oxide as catalyst, though this is probably reduced in the oil itself. *Shukoff*[2] has patented the use of *Nickel Carbonyl*, obtained by the action of carbon monoxide on reduced nickel at low temperatures. This body is soluble in oil, and the solution is afterwards heated to a temperature of 200° C. when the carbonyl is decomposed and metallic nickel set free.

Several patents are concerned with the use of carriers for the reduced nickel. Thus pumice, asbestos, kieselguhr and other porous bodies are suggested. These are impregnated with the nickel salt, and the hydroxide precipitated on them by means of alkaline hydroxide, and are subsequently reduced in hydrogen.

Recently it has been shown[3] that hydrogenation of oil can be induced without catalytic aid by the use of high temperatures and a high pressure of hydrogen.

§ 209. Recovery of Catalyst.

This matter has also been the subject of many patents. The catalyst may be recovered by filtration of the hardened oil, or by treatment with

[1] The Vereinigte Chemische Werke A.G. in their patent state that 0·001 per cent. of Palladium is effective.

[2] Ger. Pat. 241,823 of Jan. 18, 1910.

[3] Bergius, *Zeitsch. Augen Chem.*, 1914, **27**, 522.

acid and subsequent re-precipitation of the hydroxide. *Wilburschewitsch*[1] extracts the catalyst with a solvent.

§ **210.** The HYDROGEN is mainly prepared by two methods, viz.:

(1) **Electrolytic.**

This method produces a very pure gas and is eminently suitable for small installations. In the electrolytic production of caustic soda by treatment of brine a large amount of hydrogen is produced as a by-product which might be very usefully employed in this manner.

(2) **Iron-sponge steam process.**

Oxide of Iron is reduced by means of water-gas carefully freed from sulphur. Steam is then passed over the heated reduced metal, yielding the oxide and hydrogen, the former being again reduced by water-gas.

§ **211.** USES OF HARDENED FATS.

As any desired degree of hardening of a fatty oil may be obtained by regulating the **length of time** of the operation, the process is a very valuable one for the production of soap material, candle material, and edible fats.

Apart from the actual hydrogenation of the unsaturated glycerides, the objectionable odour of the fish oils and other low class fats is completely removed. This effect is however only obtained if the hydrogenation is carried almost to the full extent[2]. It is thus not possible by this method to deodorise an oil without hardening it, and the process has for this reason no interest for users of oils as such, as for instance the soft soap manufacturers.

An English firm of soapmakers[3] has patented the use of hardened fat in the manufacture of soaps. The soap made from such stock is capable of giving satisfactory results with a much larger percentage of rosin than would be the case using tallow or palm oil. Hardened fat appears to differ[4] from natural fats in the ease with which it is saponified by alkalies.

With regard to edible fats, the question of their digestibility compared with the natural product is important, and no difference has been found in this respect. A certain minute amount of the catalyst remains in the fat[5], but this is now much less than formerly. Recently[6] feeding trials with commercially hardened oils containing from ·07 to 6 milligrammes of nickel per kilo have been carried out, and no ill effects have been evidenced.

[1] Fr. Pat. 426,343.
[2] Iodine value below 50.
[3] Crossfield and Sons, Eng. Pat. 13,042 of 1907.
[4] Leimdorfer (*Seifensieder Zeitung*, 1913, **40**, 1317) suggests that "hardened" stearin is an allotropic modification of the natural glyceride.
[5] As the fatty acid salt or "soap" of the metal.
[6] Lehmann, *Chem. Zeit.* 1914, **38**, 798.

ANALYTICAL ASPECTS.

To the analytical chemist the introduction into commerce of this process has meant a landmark of the first importance. An element of doubt has been introduced into the interpretation of most of the determinations hitherto relied upon.

Generally speaking, this applies to the fats only, as (see above) little advantage is gained from partial hydrogenation. Although the detection of the presence of even minute traces of nickel is possible[1], the catalyst can now be completely removed, except in the case of oil containing free fatty acids.

With regard to the various tests, the most important is the Iodine value. This will give a direct estimate of the degree of hydrogenation. The *saponification value* will remain almost unchanged, but *Norman* and *Hugel*[2] state that hydroxyl groups are split off during the hardening process. Thus Castor Oil might not be distinguishable on complete saturation from other oils, although the *Acetyl value* would in practice only be lowered. Hardened fish and whale oils are said[3] to contain arachidic and behenic acids ; an oil adulterated with such products may be with difficulty distinguishable from Arachis Oil in the *Bellier Test*. (See p. 139.)

Of the distinguishing colour reactions, the *Halphen* test is rendered negative by the destruction of the substance producing the colour. The use of Pyridine[4] in place of amyl alcohol and a closed tube for the test has been found more sensitive and should therefore be employed in the case of a suspected hardened cottonseed oil to detect traces of the chromogenetic body.

The *Baudouin* test for sesamé oil is also negatived[5]. Fortunately the distinctive aromatic alcohols, *Phytosterol* and *Cholesterol* are unaffected, so that the animal or vegetable origin of a hardened fat is still ascertainable.

[1] See use of α-benzildioxime for detection of Nickel, vol. II. (Atack, *Chem. Zeit.* 1913, **37**, 773.)

[2] *Chem. Zeit.* 1913, **37**, 815.

[3] *Ibid.*

[4] Gastaldi, *Chem. Zeit.* 1912, **2**, 758.

[5] Some hardened sesamé oils, however, have been found to give this reaction.

CHAPTER XXI

MINERAL OILS AND WAXES

PRODUCTION AND REFINEMENT

§ 212. Petroleum and shale oil, as has already been shewn, are very complex mixtures, and in order to produce the various commercial articles obtained from these, they are submitted to various processes the objects of which are:

(i) to separate the complex mixtures into simpler mixtures;

(ii) to remove objectionable compounds from the mixture;

(iii) to transform certain constituents, not required as such, into other compounds having a greater technical and commercial value.

The actual sequence of operations carried out varies in different works and with different petroleums and shale oils, but the principles governing these operations are the same in all cases.

SEPARATION OF COMPLEX LIQUID MIXTURES INTO SIMPLER MIXTURES.

§ 213. Distillation.

The separation of the complex mixture (petroleum or shale oil) into simple mixtures is carried out by the process of *distillation*. In the simplest case distillation consists in converting a liquid into a vapour, condensing the vapour back into a liquid and collecting the condensed liquid. Distillation is said to be "destructive"—"*destructive distillation*"—when the substance undergoing the process is so heated that it is chemically decomposed into other bodies, some of which may be gases or vapours—or both—and solids, e.g. the destructive distillation of coal produces (1) coal gas, (2) liquid products obtained by cooling the vapours produced during the heating, (3) solid products produced similarly as well as a residue left in the still, viz. coke.

§ 214. In order to obtain a clear understanding of the technical processes of distillation as carried out in the petroleum and shale oil industries it is necessary to thoroughly grasp the principles underlying the process of distillation.

Liquids tend to change into vapours at all temperatures thus producing a pressure which is known as the "vapour pressure"; this vapour pressure varies with the temperature, being greater at higher temperatures than low. For all liquids there is a *temperature* at which the *vapour pressure* of the liquid is *equal* to the *pressure of the superincumbent atmosphere*; at this temperature the liquid changes into vapour with *ebullition*, i.e. the liquid boils. Since a liquid exerts a vapour pressure at all temperatures it is evident that a liquid can be made to boil, i.e. change into a vapour with ebullition, by reducing the pressure of the superincumbent atmosphere until it reaches the pressure

LABORATORY APPARATUS FOR DISTILLATION UNDER REDUCED PRESSURE

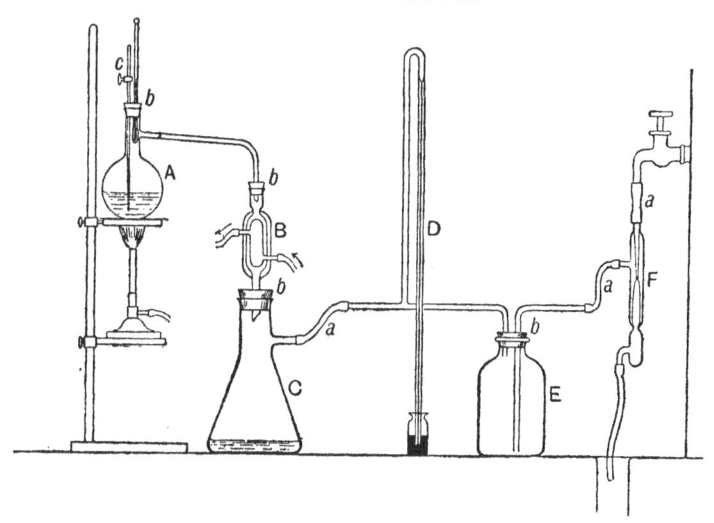

Fig. 19. A. Distillation flask fitted with an air inlet *c* and thermometer.
B. Double surface condenser.
C. Strong flask to receive distillate.
D. Mercury manometer.
E. Strong bottle to regulate changes of pressure.
F. Water-pump.
a, a, a. Stout rubber pressure tubing.
b, b, b, b. Good rubber corks.

of the vapour of the liquid at the particular temperature chosen; by such reduction of the atmospheric pressure a liquid can be made to boil at any desired temperature—with certain restrictions—and hence can be distilled at any desired temperature; when distillation takes place under reduced atmospheric pressure it is termed "distillation in vacuo." Fig. 19 represents a laboratory apparatus for distillation under reduced pressure.

§ **215.** Since petroleum and shale oil are complex mixtures containing definite liquid compounds each possessing a definite boiling point, the question arises, what will be the boiling point of such a mixture? Experience shows that these bodies possess no definite boiling point and hence on distillation no definite constituent is obtained but a mixture differing in composition from the original mixture.

The question of the distillation of liquid mixtures whose constituents are miscible in all proportions has been fully studied by Konowalow with the following results:

1. *Mixtures whose constituents can be separated more or less completely by fractional distillation.*

In this class of mixtures it has been found that the vapour pressures of the mixtures vary increasingly with the temperature:

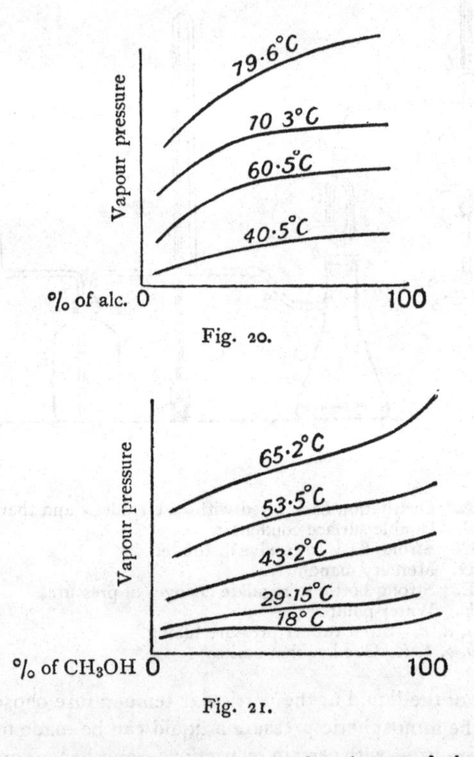

Fig. 20.

Fig. 21.

Figs. 20, 21 give the vapour pressure curves for mixtures of ethyl alcohol with water, and of methyl alcohol with water. Now since the boiling point varies inversely as the vapour pressure, those mixtures with highest vapour pressure will have the lowest boiling point, and hence from above curves it will be seen that the mixture containing the greatest percentage of alcohol will boil first, and hence in distillation the distillate will contain at first a mixture of alcohol and water richer in alcohol than the original mixture; by collecting the distillate in several portions or fractions of the original liquid, different mixtures will be obtained. On submitting each one of these fractions again to distillation and collecting the distillates again in portions it is possible to obtain nearly a complete separation of the alcohol from the water; it will be noted that as distillation proceeds in each case the boiling point of the liquid being distilled slowly rises, and hence the process of separation can be followed by taking the temperature of the vapour in the distillation vessel. All mixtures whose vapour pressure curves are similar to Figs. 20, 21 behave in a similar manner on distillation.

2. *Mixtures with constant boiling-point.*

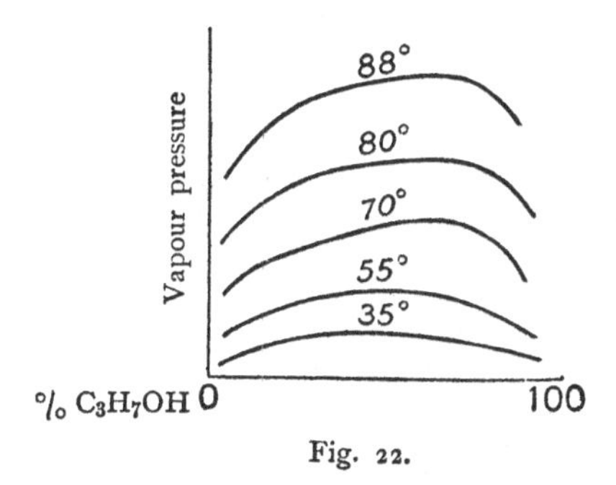

Fig. 22.

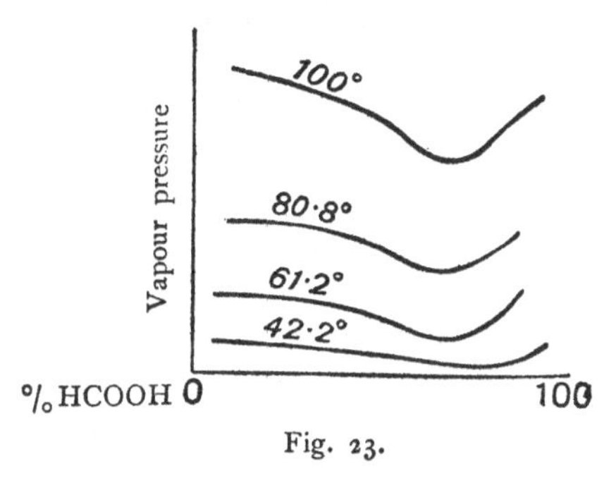

Fig. 23.

Figs. 22, 23 give the vapour pressure curves of propyl alcohol with water, and of formic acid with water.

It will be noticed that in the case of propyl alcohol and water that at each temperature there is a mixture which possesses a maximum vapour pressure and hence a minimum boiling point: this mixture contains about 70 % of propyl alcohol; hence on fractionally distilling any mixture of these two we finally obtain a mixture whose boiling point is constant and containing about 70 % of propyl alcohol. In the case of formic acid the mixture containing about 75 % of formic acid possesses the lowest vapour pressure at all temperatures; hence on distilling mixtures of formic acid and water this constant boiling point mixture tends to collect in the still.

§ 216. Steam distillation.

The vapour pressure of a mixture of two immiscible liquids is the sum of the vapour pressures of each constituent. On distilling such a mixture, when the sum of the vapour pressures of the two liquids equals the atmospheric pressure, the liquids distil over and the distillate contains each of the constituents such that its composition remains constant until one of the constituents has completely distilled over.

The composition of the distillate can be calculated when the following data are known: (1) the boiling point of the mixture, which is always lower than that of either constituent, (2) the vapour pressure of each constituent at the boiling point of the mixture, e.g. a mixture of water and n. octane has a boiling point of 89·5° C., and the vapour pressure of water at 89·5° C. is 515·6 mm., whilst the vapour pressure of n. octane at 89·5° C. is 244·4 mm. The molecular weight of water is 18 and the molecular weight of n. octane is 114, hence the distillate will contain water and n. octane in the proportion of

$$18 \times 515{\cdot}6 \text{ is to } 114 \times 244{\cdot}4, \text{ i.e. } 9280 \text{ is to } 27861, \text{ i.e. } 1 \text{ is to } 3.$$

It is thus seen that if a substance has a high molecular weight as well as a high boiling point it can be distilled at a lower temperature than its normal boiling point by blowing steam into it: this property is made use of in refining petroleum and shale oils; in order not to decompose the higher boiling point members during distillation, steam is passed into the stills, when these members distil over with the water at a lower temperature than they would normally.

Fig. 24 depicts a laboratory method of carrying out a steam distillation.

§ 217. Distillation in vacuo.

As pointed out in § 214 the boiling point of a liquid can be lowered by reducing the atmospheric pressure in the still; this property is also made use of in distilling those substances which would undergo decomposition at their boiling point, or possess a very high boiling point; as e.g. the lubricating oils and some mineral waxes. Fig. 19 shows a laboratory arrangement for distillation in vacuo.

§ 218. Destructive Distillation.

In the mineral oil industry two distinct types of destructive distillation are carried out represented by:

1. **The Decomposition of Petroleum and Shale Oils** during the distillation. The amount of decomposition depends upon the degree of heat to which the oil is submitted and can be controlled within certain limits. At high temperatures the higher paraffins are decomposed into lower members with the simultaneous formation of Olefines, e.g.

$$\underset{\text{Paraffin}}{C_{2x}H_{4x+2}} = \underset{\text{Paraffin}}{C_xH_{2x+2}} + \underset{\text{Olefin}}{C_xH_{2x}}.$$

The lower paraffin can then undergo a similar decomposition, and this process will continue until a paraffin is formed which boils at prevailing temperature without decomposition. The olefines formed probably undergo polymerization forming members of higher molecular weight, viz.

$$2\,C_n H_{2n} = C_{2n} H_{4n}, \text{ etc.}$$

It should be noted that this process of destructive distillation or "cracking," as it is technically termed, gives a larger yield of the lighter and burning oils at the expense of the lubricating oils and paraffin wax. Consequently the distiller regulates the distillation by admitting more or less steam to the still, so as to produce more or less of the fractions required.

LABORATORY APPARATUS FOR STEAM DISTILLATION

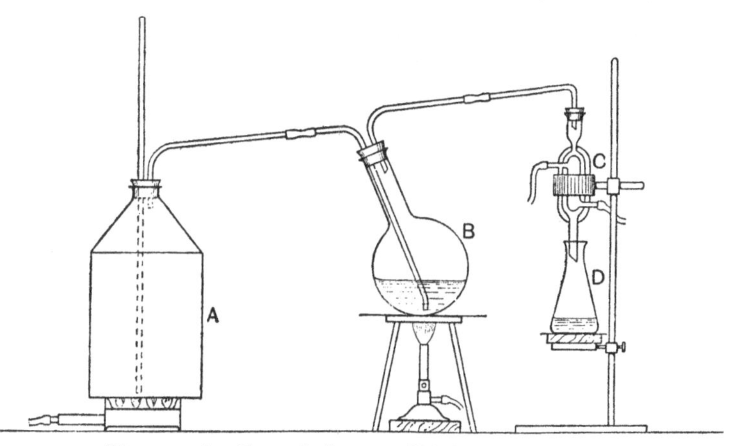

Fig. 24. A. Taper tin (1 to 2 galls.) for producing steam.
B. Flask containing liquid to be distilled.
C. Condenser.
D. Receiver.

Cracking is produced technically by

(1) Distilling under increased pressure by throttling the outlet of the still.

(2) Allowing the condensed distillate to fall back on the highly heated charge in the still.

The product obtained is not identical with the natural distillate, and is much inferior, taking more acid in subsequent refining and losing its colour on keeping. The odour is also much stronger.

2. The Decomposition of Shale and Coal.

The distillation of these bodies consists in the heating of them to more or less high temperatures in the absence of air. Depending upon the temperatures used and the nature of the coal or shale, so the products of distillation vary. With each substance more or less gas is obtained and liquid distillates, viz. coal tar oil and shale oil, whilst coke is left in the retorts.

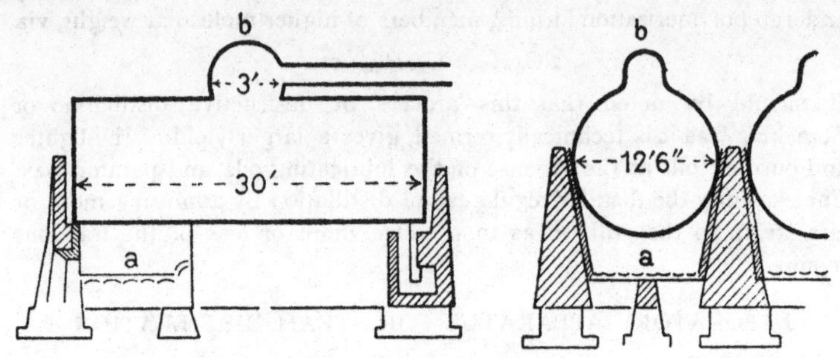

Fig. 25. One of a bench of cylinder stills. There are usually six in a bench, capacity 600 barrels of crude oil each. Constructed of wrought iron or steel, and formerly completely bricked in, but the top is now commonly left exposed to obtain "cracking" of the distillate.
a. Grate.
b. Dome with outlet to condensers.

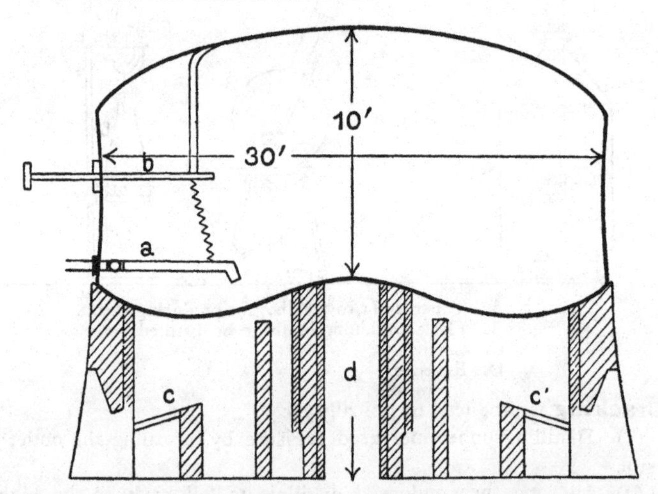

Fig. 26. Section of "Cheese Box" type of still. It is supported on arches of brickwork and has 17 fireplaces. The discharge pipe (*a*) is capable of being raised and lowered by means of (*b*).
Fireplaces at *c* and *c*′. (*d*) is central flue leading to chimney.
Capacity 1200 barrels crude oil.

§ 219. Types of Distillation Plant.

Figs. 25 and 26 show two models of stills in use for the distillation of crude petroleum. The approximate dimensions and the capacity are also shown.

In Russia a continuous system of distillation is largely employed. Each unit consists of 14 to 16 stills which are so placed that the oil flows through the series by gravitation. The method is shown in Fig. 27. The

DIAGRAM OF CONTINUOUS DISTILLATION PROCESS

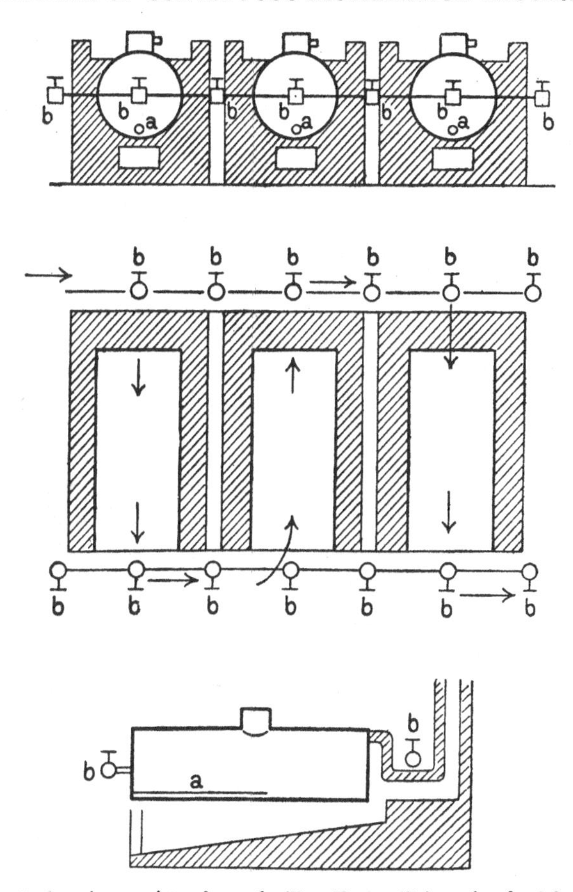

Fig. 27. Each series consists of 14–16 stills. Each still is 24 feet by 8 feet. The oil enters and flows out by means of 4 inch pipes connecting with 6 inch mains. The temperature is gradually increased in each still of the series. The flow is by gravity and once adjusted the whole series needs only a single man's attention. Stills when out of repair can be disconnected from the series. The fuel used is liquid astatki "atomised" by means of a steam jet.

oil enters and flows from each still through 4-inch pipes controlled by valves and connected with 6-inch mains. The temperature is increased in each still of the series. There is a great economy of labour and handling, as the stills once adjusted are capable of being worked by one man. When a still requires repair it can be disconnected from the series.

The fuel employed in Russia is the residue from the distillation, known as "*astatki.*" This is conveyed hot in pipes to the furnaces, and the fuel is "atomised" by the action of a jet of steam, by means of which a perfect combustion takes place.

The *condensers* are mainly of the parallel tube type. These consist of pipes cooled in wooden troughs with circulating water and laid with a slight fall. The uncondensable *gas* is removed by means of a vertical pipe and trap. After condensation the oil passes a glass observation door, where the colour, etc. can be seen and then connects with 5 or more pipe junctions, with valves, by means of which the distillate may be run into any one of a number of underground tanks (Fig. 28).

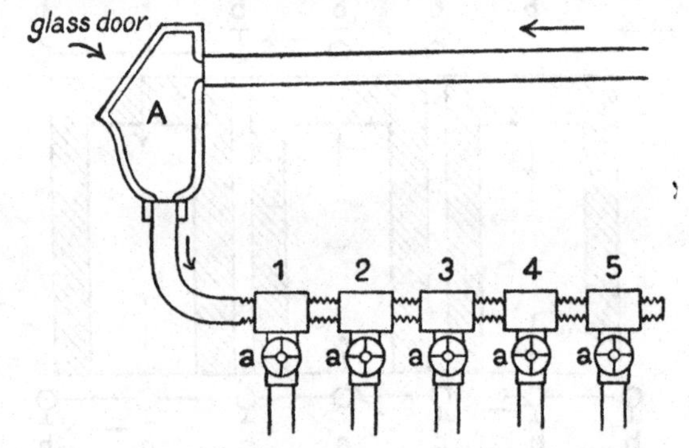

Fig. 28. Distributor for separating the various fractions. The discharge pipes 1—5 connect with underground tanks.
At *A* is the observation window where samples are taken and the colour and consistency of the distillate may be judged. The controlling valves for the various tanks are at *a*.

§ **220. Character of the Products.** The products of distillation vary in character and amount according to the source of the crude petroleum. Thus American petroleum differs from Russian in several important particulars. This is shown below; the figures representing percentage yields.

	United States	Baku
Light petroleum	16·5	2–5
Kerosene	50–54	30 [+from 4–6 "solar oil"]
Lubricating oil	17	37+28 "light lubricating solar oil"
Paraffin wax	2	——
Residue	loss 10 coke	24 astatki

The residuum in Russia being employed as a fuel is in good demand and there is thus less inducement to "crack" in distillation. Compared with American oil, the yield of light naphthas is small, but there is a high proportion of lubricating oil, and much valuable fuel material (astatki).

In the first distillation the American practice is to separate the products into 3 main divisions:

(1) Fraction B.P. up to 150° C. (crude naphtha).
(2) „ „ „ 150°–300° C. (crude kerosene).
(3) „ „ above 300° C. left in still.

These fractions are each redistilled as described under, and separately refined. The heavier fractions of (1) will be similar in character to the lighter fractions of (2), and a quantity of heavy oils of the lubricating class will be obtained on redistillation and fractionation from the crude kerosene fraction.

In Russia much the same method is used, but the still residue contains a large amount of the valuable liquid fuel "astatki," and distillation of this fraction to coking point is not found remunerative.

SHALE.

§ **221. Shale** differs from petroleum in many important respects, but chiefly in that the oil is *not obtainable by ordinary distillation*, or by extraction. In order to obtain oil the shale must be destructively distilled and the complex carbon compounds which it contains split up. In this respect it resembles coal but differs from the latter in yielding hydrocarbons belonging, not to the benzene series as in the case of coal, but to the paraffin and olefin groups like petroleum. It also resembles coal in yielding in addition to oil, a watery liquor rich in ammonia, from which sulphate of ammonia is produced. The yield of this substance often compensates for a poor yield of oil in the lower class shales.

§ **222.** Deposits of shale exist in the following countries:

Scotland,
France,
New South Wales,
Nova Scotia,
Servia,
Saxony.

The Scotch Industry is the most important, and the French is the oldest, but in the latter country the oil has not been able to compete with Petroleum.

Oil Shale is dark grey-black material with a laminated horny fracture. Specific gravity 1·75. The mineral constituents vary up to 75—80 per cent., and the yield of crude oil varies from 18—40 gallons per ton. The *crude oil*, or "once run oil," is dark green, of specific gravity 0·86—0·89, and solidifies at 90° F., containing 10—15 per cent. of paraffin wax. It consists chiefly of Paraffins and Olefins, but contains 1—1·5 per cent. of Nitrogen in the form of Pyridine bases.

A very rich shale was the material known as "Boghead coal." This yielded 70 per cent. of volatile substances, and from 120 to 130 gallons of crude oil per ton of material. The deposit was worked out in 1862. The richest shales at the present time give a yield of about 36—40 gallons per ton of crude oil and 25 to 35 lbs. of ammonium sulphate. The area of the deposits is usually small and the seams average four feet in thickness.

§ **223.** The shale is usually **mined** by blasting, and the mineral conveyed by chain haulage to the retorts. A toothed breaker reduces the lumps to pieces of uniform size, about ½ foot square. The cost of shale delivered to works is stated to be 4 to 6 shillings per ton.

DIAGRAM OF A SHALE RETORT IN SECTION

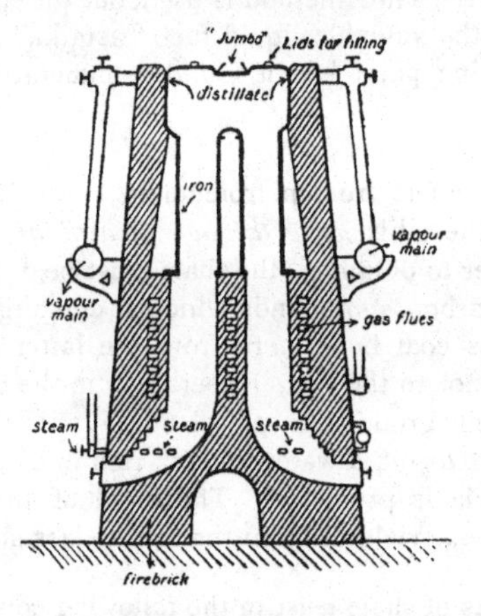

Fig. 29. Diagram of the **"Pentland Retort"** (Young and Beilby, 1882). The top section is of iron, the lower of firebrick. The chambers of the retorts are in groups of four surmounted by a "Jumbo." The lower part is filled with spent shale and heated. The upper part is charged with raw shale and steam blown in at the bottom. The products are drawn off through the vapour mains. The spent shale is removed below and the charge gradually subsides in the retort, fresh material being introduced at the top.

Capacity = 30 cwt. per retort ; 8 cwt. every six hours. A bench consists of 80 retorts, which deal with 100 to 120 tons of shale per day.

§ **224. Distillation** was first carried out in coal gas retorts but the vertical type was soon introduced. Steam was used to carry off the vapours. The spent shale is now used as fuel in the furnace, and the gas produced is also employed as a heating agent. The temperature employed is about 800° F. The modern form of the retort is shown in Fig. 29 together with a description of its working. Condensation is carried out in a similar manner to petroleum. The gas is "scrubbed" to remove tarry matters, and is then chiefly used for fuel in the retorts.

OZOKERIT.

§ **225.** As mentioned on page 194 ozokerit has been found in a meteorite. It is found very widely distributed, but chiefly in *Galicia*[1]. It is probably an oxidation and condensation product of petroleum. The industry dates from 1860 and is chiefly carried on in the neighbourhood of Boryslaw. The production was highest in the year 1885 (13,000 tons) and has since fallen to a fraction of that quantity.

It was formerly mined on a very primitive system, but modern methods are now in vogue in many places. The products obtained are either the nearly pure wax in fragments, or earth containing much ozokerite. The pieces of wax are melted in water; the earthy fragments are sorted and extracted with benzene.

LIGNITE OR BROWN COAL.

§ **226.** This is an important industry of Saxony. The brown coal varies widely both in appearance and composition. Many varieties show a woody structure. On distillation it yields acetic acid and is thus distinguished from shale and coal which give ammoniacal watery distillates. The crude tar is light and of a buttery consistency. The best coals yield 2—10 per cent. tar, 25—60 per cent. coke, 20—40 per cent. aqueous layer, and 5—25 per cent. gas. The following are the average yields obtained in percentages of the tar:

		° F.	Specific Gravity
Naphtha (1)	2—6	130°—170°	0·71—0·75
Naphtha (2)	8—30	170°—220°	0·75—0·82
Solar Oil	15—50	220°—290°	0·82—0·88
Lubricating Oil	15—30	290°—320°	0·88—0·90
Paraffin Wax	1·5—12	above 320°	0·89—0·91
Asphalt	14—20		
Gases	3—10		
Water	8—20		

In place of distilling lignite, it may be treated for a high melting point wax. This is performed by either extracting with benzine or distillation with superheated steam, and the hard crude wax so obtained refined as indicated on page 191. The yield of crude bitumen wax is 7—10 per cent. of melting point 70°—85° C. It is a lustrous, brittle, non-crystalline substance of dark brown colour.

REFINEMENT OF MINERAL OILS AND WAXES

§ **227.** The refinement of the distillates derived from Petroleum, Shale and Lignite all follows the same course. The first treatment is designed to separate the constituent hydrocarbons of the crude fraction more completely. This is done by re-distilling, and collecting the distillate in fractions, between definite boiling points or specific gravities. The second treatment is necessary to remove the substances present other than hydrocarbons, such as phenols, pyridine bases, and tarry

[1] Other countries where ozokerit occurs are Servia, Egypt, Russia, Orange River Colony, Utah.

matters which colour the oil. This is accomplished by treatment with a specified quantity of strong sulphuric acid followed by a wash, and in some cases an after treatment with weak alkali. The more volatile the fraction, the easier, generally speaking, is it to refine and obtain colourless. More volatile fractions require less sulphuric acid and a shorter time of treatment.

§ 228. Naphtha.

Naphtha, or the crude light fraction from various crude oils, is fractionated into various grades which have different names in different countries.

Typical fractions are given on page 198. From $\frac{1}{4}$ to $\frac{1}{2}$ per cent. of sulphuric acid of 66° Bé is used, and it often has a final wash with weak caustic soda.

The yields of naphtha vary, being highest in some East Indian oils, moderate in American and low in Russian. Shale yields roughly about half the quantity of average American petroleum.

§ 229. Kerosene.

Kerosene, or illuminating oil, the middle fraction from crude oils, is treated with about $1\frac{1}{2}$ per cent. of 66° Bé sulphuric acid in vessels of about 1,200 barrels capacity (48,000 galls.). It is previously freed from water by settlement, and the acid is admitted through a spray and thoroughly mixed, either with mechanical agitation, or by means of an air blast. The acid is frequently added in three portions; a small quantity first to remove traces of water, and two further additions, which are allowed to act for one hour and then settled and drawn off. The acid treatment is followed by a washing with water, which is allowed to fall in a shower through the oil, any agitation being apt to cause troublesome emulsions. This is succeeded by an agitation with a one per cent. CAUSTIC SODA SOLUTION, after which the oil is again washed with water. In order to reduce the flash point, the oil is "sprayed"—if necessary—i.e. forced through fine spray nozzles so as to induce spontaneous evaporation of the very light fractions.

§ 230. Lubricating Oil.

The **Lubricating Oil** crude fraction is further fractionated in a vacuum still, under reduced pressure with superheated steam. The object of this is to prevent "cracking" of the oil into lighter fractions, and to assist the carrying over of the heavy vapours. The loss, in treatment with the gas and coke amounts to about 12 per cent. The condensers are kept warm to prevent the solidification of the paraffin wax (U.S.A.).

The first fraction is known as SOLAR OIL (see also page 199) also called "intermediate" oil. This is too light for lubrication, and also too heavy for burning, but finds an outlet in agriculture and other purposes.

The next fraction is the SPINDLE OIL, also graded as "heavy" and "light," used for lubrication of fine machinery, dynamos and high-speed work. It is frequently blended with sperm oil for this purpose. The following fractions are of ENGINE OIL grade and intermediate in character. The best oils are characterised by an absence of paraffin wax and have low freezing points. Finally come the CYLINDER OILS of varying degrees of viscosity, and here again it is very necessary to free the oils from paraffin wax. In the case of the Russian oils, the distillates are practically free from wax. (For details of the treatment of the oils'

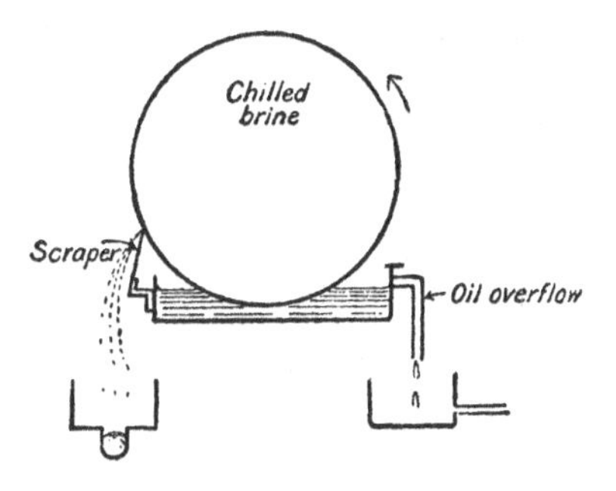

DIAGRAM OF SIMPLE FORM OF WAX SEPARATOR

Fig. 30. The iron cylinder in which chilled brine circulates is slowly revolved in a shallow trough of oil. The wax solidifies and adheres to the cylinder while the oil runs off. A scraper removes the wax to a draining trough.

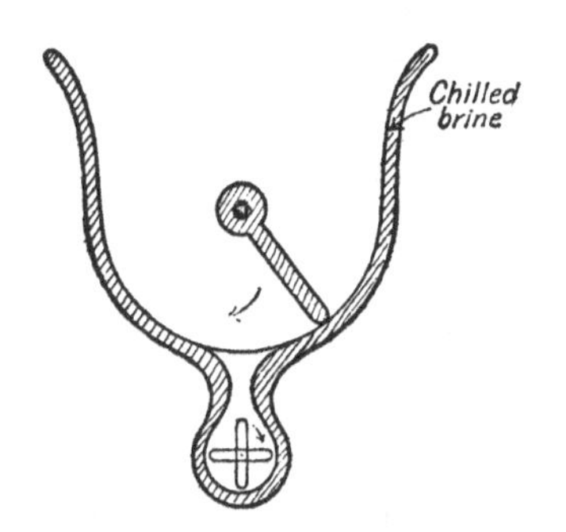

DIAGRAM OF HENDERSON'S COOLER

Fig. 31. Consists of a trough shaped vessel, with a jacket in which chilled brine circulates. In this case the scraper revolves inside the vessel, and deposits the wax in a well, in which it is kept stirred and pumped to filter presses which remove remainder of oil.

heavier fractions for removal of wax, see below.) The **acid treatment** consists in agitation with 5 to 10 per cent. of sulphuric acid of 66° Bé. at a temperature of 80°—100° C. The "acid tar" ("foots") is drawn off and the oil treated with weak soda, and finally washed with water at 60°—70° C. Mechanical agitation is preferred to the use of an air blast. Emulsions are easily produced, and are very difficult to separate. The yield of lubricating oil varies very much with the source of the crude oil. Russian Oil gives the highest yield, shale oil and some lignite crudes come next, and American varieties on the average have the lowest yields.

Mineral Waxes.

§ **231. Paraffin Wax** is obtained chiefly from shale oil, lignite, and American and East Indian petroleum. The last named yields a very high melting point product. Russian petroleum contains practically no wax. The highest yields are obtained from shale oil, which contains upwards of 10 per cent. of wax. The lubricating fractions are cooled with freezing brine and the wax separated in the manner shown in the diagrams (figs. 30 and 31).

The crude wax, known as SCALE, is then refined by what is termed **sweating**. This consists in exposing the solidified wax to a

DIAGRAM OF SWEATING PROCESS

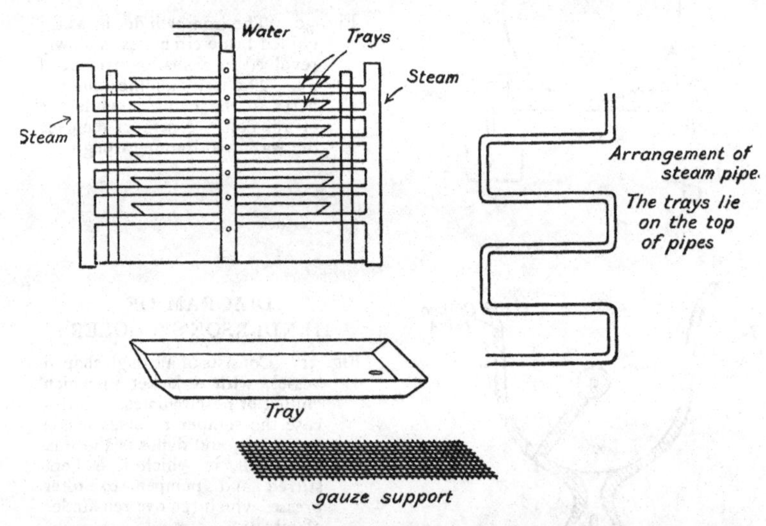

Fig. 32. The Henderson process of "sweating" consists in the use of trays, supported on stands and lying over steam pipes as shown. Each tray has an outlet connected with discharge tubes and a false bottom of gauze (sixteen meshes to the inch). Before charging the trays are filled with water up to the gauze support, the trays are then filled with melted wax, and on setting of the latter the water is run off, leaving a cake of wax supported on the gauze. The stove is heated to 80° C. for some time and then maintained at the sweating temperature until all the liquid products have run off; when the steam pipes are heated and the melted wax run off.

warm temperature in shallow cakes, when the liquid products gradually "sweat" out and are carried away. The melted wax is finally agitated for 1½ hours with animal charcoal or siliceous earths, filtered and moulded into cakes. Fig. 32 is a diagram and description of the arrangements used for "sweating."

§ 232. **Ozokerit Wax.**

The refining of **ozokerit** is carried on in refineries at Hamburg, Drohobycz and Vienna. The refined product is known as "ceresin." It was formerly refined by distillation with superheated steam[1]. The modern process is due to

DIAGRAM OF CONTINUOUS TYPE OF EXTRACTION APPARATUS

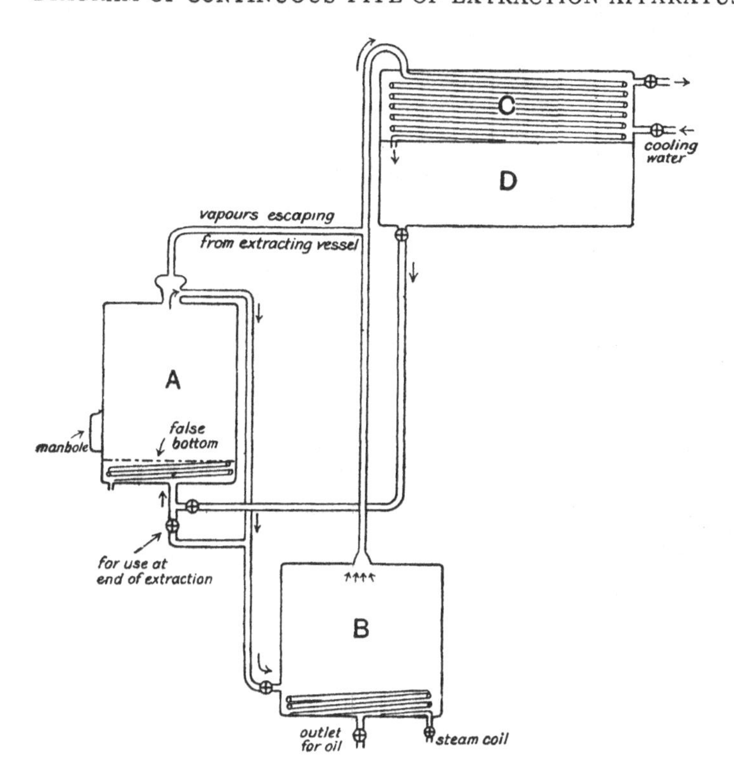

Fig. 33. A. Extracting vessel with manhole for charging, false bottom, and steam coil.

B. Evaporating vessel. The solvent is here distilled constantly off, leaving the oil and the vapour condenses in

C. The condenser, and runs into

D. The store tank, whence it continuously flows into the extracting vessel A.

Von Boyen. Open wrought iron pots are employed of capacity about 3000 kilos. A hood is provided for the fumes produced. The wax is melted and heated for 4–6 hours till anhydrous. At about 120° C. 18 to 20 per cent. of FUMING SULPHURIC ACID (78 per cent. anhydride) is added slowly during vigorous stirring

[1] Field's patent, 1870.

with an iron paddle. Copious fumes of sulphur dioxide are evolved. The fire is then increased, and at 140° C. a precipitate ("acid asphalt") forms which separates from the wax. As the temperature continues to rise the precipitate becomes granular, and the maximum is reached at 175° C., at which it is kept for 3 hours. The next day the wax is run off from the settlement, de-acidifying powders are added, and much frothing occurs. The charge is raised again to 130° C. Filtered samples should now be pale yellow in colour. Decolourising powder is added[1] and the hot wax is filtered.

Another process consists in melting the crude wax with a large amount of animal "char," and extracting with benzene. The extraction plant used is shown in Fig. 33. The product obtained is stated to be more plastic than the acid refined wax.

§ 233. Yields of Fractions from Distillation.

The following table gives the average yields of commercial fractions by the distillation of petroleums, etc.

	U.S.A.	Russia	Crude Oil Shale	Crude Oil Lignite
Naphthas ...	17	2	6	2–17
Illuminating Oil (Kerosene)	50–54	36	32	—
Lubricating Oils	17	62 (with solar oil)	24	18–40
Paraffin Wax	2	—	13	5–40
Loss	10	0[2]	25	to 40

[1] Precipitated silica and aluminium silicate have been found superior to animal charcoal for this purpose.

[2] Distillation not carried to coking point, see § 220.

SECTION VI

OLEO-RESINS AND ESSENTIAL OILS

§ **234.** As stated in a former chapter (page 3) the essential oils form a class distinct in themselves, being of fundamentally different chemical composition and having quite a different application technically from the previously described oils.

There is however one natural product belonging to this class which it is customary to describe along with the "fixed" oils and fats. This is the oleo-resin from which common "Rosin" (Colophony) and "Turpentine" are obtained. The reason for this is found in the fact that both these products are used very largely in the industries usually associated with the fixed oils and waxes, viz. soap-making and polish-manufacture (boot polish, furniture polish, etc.), whilst the other and less common members of the class are employed exclusively in varnish-making, perfumery and allied trades. It is therefore thought desirable to include a brief description of these two products.

CHAPTER XXII

TURPENTINE AND ITS SUBSTITUTES

§ **235.** "Turpentine," or "Oil of Turpentine" is obtained by the distillation of the exudation from various species of trees belonging to the *Coniferæ*. The industry is carried on principally in the United States and in France, but smaller quantities are also produced in Germany, Russia, Spain, Sweden and India.

The principal varieties of trees are as follows:

America—	Pinus	Australis.	German—	Pinus	Sylvestris.
France—	"	Pinaster.	Russia—	"	"
	"	Maritima.	India—	"	Longifolia.

§ **236.** The method used for obtaining the natural product is to "box" the trees during winter by making incisions into the trunks about 1–2 feet from the ground. These incisions are in the form of cups or "boxes" which will hold about ½ to 1 gallon of the exudation. The trunk above the "box" is cleft in several places so as to assist the flow of the oleo-resin, which takes place chiefly from March until September. The exudation is then collected from each tree, and conveyed to the refinery where it is placed in stills. In the old works these are fire heated, but more modern plants employ superheated steam for the distillation. The turpentine distils over, and is condensed together with water vapours in a water condenser, while the residue in the still consists of the rosin of commerce. The yield of turpentine is about 20 per cent. of the charge. The use of fire heat produces a darker coloured rosin than that obtained with superheated steam owing to overheating of portions of the still-charge. This is mitigated by the use of a steam pipe blowing steam through the heated mass, and by this means the rate of distillation of the turpentine is also considerably increased.

The distillate consists of the turpentine and condensed steam. After settlement it is transferred to barrels for shipment.

§ 237. RUSSIAN TURPENTINE.

Russian turpentine is obtained by distillation of the wood of the Scotch fir (*Pinus sylvestris*). It has a strongly pronounced odour which unfits it for use with varnishes, etc.

Many patents have been taken out for its deodorisation, and though some are effective the unpleasant odour has a tendency to return after exposure to air.

Chemically it contains a very much smaller percentage of *Pinene* than American turpentine.

For the recognition of Russian turpentine in American the following test[1] is applied :

10 drops of Hydrochloric acid are added to a mixture of 5 c.c. of the suspected oil and 5 c.c. of Acetic Anhydride. The mixture is cooled and a further 5 drops of Hydrochloric acid is added, and it is again shaken and cooled. In the presence of Russian Turpentine there is considerable darkening, but if pure American, only a faint yellow colour is developed.

§ 238. WOOD TURPENTINE.

This is a product obtained in a similar manner to Russian turpentine, by distillation of the branches and roots of the pine. Owing to the improved methods of distillation--with steam—and fractionation of the distillate, a product is now obtained which resembles a genuine turpentine so closely as to be differentiated therefrom with difficulty[2]. It should be carefully tested for smell by exposing a portion in a dish together with some turpentine of undoubted genuineness for several hours and comparing the odour of each. Another excellent test is to shake the oil with an equal volume of SULPHUROUS ACID of 1·03 specific gravity. On separation, the "wood oil" sample will be *coloured green*. For the detection of admixtures of "wood oil" with genuine turpentine, the sample should be steam distilled, using a control experiment under identical conditions with genuine turpentine and the fractions carefully compared.

[1] Piest, *Chem. Zeit.*, 1912, 198.

[2] The Iodine value is usually lower; i.e. not over 300, comparing with genuine turpentine 380.

§ 239. Properties of Turpentine.

A colourless, mobile liquid with a characteristic odour. American is the sweetest, Russian being usually somewhat rank.

Specific gravity 0·855—0·870 at 15·5° C. (Russian to ·875).

B.P. about 160°.

Flash point (Abel) 93°—95° C.

Optical Activity:

American	+ 1° to + 15°	(dextro-rotary).
French	− 18° to − 40°	(laevo-rotary).
Russian	dextro-rotary.	

Refractive Index:

1·464—1·474 (15° C.).

7°—10° on Jean's Oleo-refractometer [1].

§ 240. Chemical Composition.

Turpentine consists almost entirely of the Terpene "*Pinene*," a hydrocarbon of the formula $C_{10}H_{16}$, which exists in two modifications α and β Pinene. The amounts of each in the American, French and German oils are as follows [2]:

	α Pinene	β Pinene
American (per cent.)	72	28
French („)	63	37
German („)	73	27

Russian and Swedish turpentine consist mainly of the terpenes, *cinene* and *sylvestrine*, hydrocarbons of the formula $C_{10}H_{16}$, and very small quantities of pinene.

The pinenes are complex hydrocarbons whose structure has been determined as the result of much work. The following structural formulae are those now usually accepted:

```
            α pinene                               β pinene
CH₂————————————CH————┐              ┌————CH———————————CH₂
|              |     |              |     |             |
|              C(CH₃)₂CH₂           CH₂   C(CH₃)₂       |
|              |     |              |     |             |
CH : C(CH₃) . CH—————┘              └————CH . C : CH₂ . CH₂
```

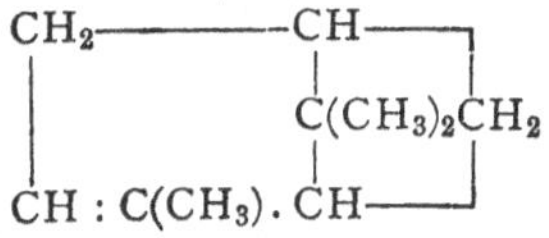

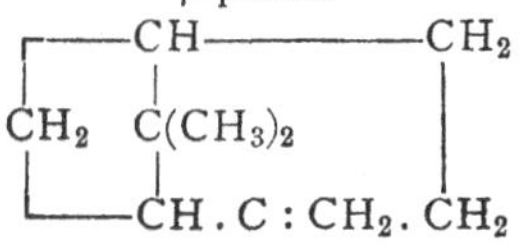

Cinene is the optically inactive form of *limonene* or *dipentene* and has the following structure:

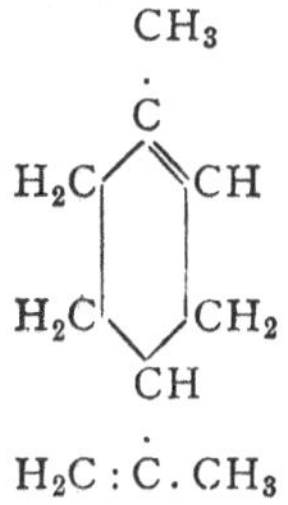

```
        CH₃
         ·
         C
      ／  ＼\
   H₂C       CH
    |         |
    |         |
   H₂C       CH₂
      ＼   ／
        CH
         ·
   H₂C : C . CH₃
```

[1] Fryer.   [2] G. Vavon, *Comptes Rend.* 1910. 150, 1127—1130.

On reduction with Hydrogen and a catalyst a hydrocarbon $C_{10}H_{18}$ is in all cases produced.

§ 241. Adulteration.

PETROLEUM DISTILLATES are indicated by a low flash point, low specific gravity and refractive index. They are determined by treatment with Sulphuric acid and subsequent distillation. In these conditions the turpentine is is converted into a viscid oil, whilst any petroleum fraction remains unchanged. The method[1] is as follows:

Place in separating funnel 300 c.c. of Turpentine to be tested.

Add 100 c.c. of Sulphuric acid (66 °/° H_2SO_4 by volume). Shake cautiously, cooling if much heat is generated. The viscid product is separated off and distilled with steam.

The oily portion of the distillate is again treated with acid (75 °/° H_2SO_4 by volume) and again distilled with steam. If volume of distillate exceeds 15 c.c. the excess is reckoned as roughly due to added petroleum distillates[2].

ROSIN SPIRIT.

Is indicated by the odour, and the behaviour on distillation. In the case of added Rosin spirit, the temperature gradually rises, and the quantity of the fraction boiling at 156°—160° C. is reduced.

"REGENERATED" TURPENTINE.

This is a product of synthetic camphor manufacture, and is used to adulterate genuine turpentine. It is detected by its high boiling point. Genuine turpentine distils at from 155° to 175° C., about 80 per cent. coming over below 165° C., whilst "regenerated" Turpentine only begins to distil at 170° C.[3].

WOOD TURPENTINE (see § 238) cannot be strictly called an adulterant, since its chemical composition is almost identical with the oil distilled from the oleo-resin It can be usually detected by its "empyreumatic" odour, due to products of decomposition derived from the wood.

§ 242. Uses.

Turpentine is largely employed as a solvent and vehicle in varnishes, polishes (where it dissolves wax) and many other manufactured products.

It is readily oxidised in contact with air, and the oxidised product, especially that obtained by blowing air through Russian turpentine, has valuable antiseptic properties.

[1] An improved method is described by Krieger in *Chem. Zeit.*, 1916, **40**, 472—473.

[2] See also in Vol. II. (Practical Work).

[3] J. Marcusson, *Chem. Zeit.*, 1909, **33**, 966—967.

CHAPTER XXIII

ROSIN

§ **243.** As previously stated (page 252) rosin is the *residue* left in the still after all the oil of turpentine has passed over. The molten rosin is run into casks with or without filtration, and here it solidifies to a vitreous mass. The colour of rosin depends on the manner in which the distillation has been carried out. It varies from "water-white" ("W.W") and "W.G" ("window glass") to nearly black. The commercial grades are represented by letters, the quality improving as one descends the alphabet. Thus "A" is the commonest quality of all, "K" is a medium quality and so forth.

Rosin is a vitreous substance, the best varieties being almost transparent and nearly colourless. It is extremely brittle. The melting point varies and is not clearly defined, the rosin often softening at a temperature below that of its actual melting point, which is usually over 100° C. The odour of rosin is pleasantly aromatic and quite characteristic.

§ 244. Physical and Chemical Data.

Specific gravity at 15·5° C.	—	1·067—1·081.	
Melting point	—	over 100° C.	
Acid value	—	130—181	Average 164
Saponification value	—	147—183	„ 175
Ester value	—	5—36	„ 15
Iodine value	—	55—180	„ 110
Unsaponifiable value	—	5—15	„ 7·5

The Iodine Value is indefinite and varies according to the time of action of the solution, degree of concentration, and amount of excess of reagent present. It acts by substitution as well as addition (see page 79).

§ 245. Chemical Composition.

Rosin consists of free acids and small quantities of anhydrides[1]. According to *Fahrion*[2] the acid (**Abietic acid**) has the following formula:

[1] Lewkowitsch, *Oils and Fats*, Vol. 1. p. 621.
[2] *Chem. Revue*, 1911, 239.

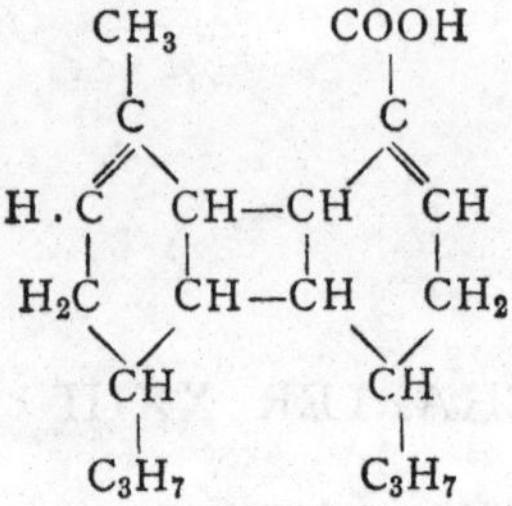

```
       CH3        COOH
        |           |
        C           C
      //  \       /  \\
  H.C     CH—CH      CH
     |     |    |     |
   H2C    CH—CH      CH2
       \  /      \  /
        CH        CH
        |         |
       C3H7      C3H7
```

It contains (like Linolic acid) two pairs of doubly-linked carbon atoms, and one acid (carboxyl) group. If it absorbed ICl regularly like Linolic acid it should give 168·2 as Iodine value, but the full amount does not appear to be taken up. On oxidation with alkaline permanganate, rosin adds on four hydroxyl groups[1] (see Linolic acid, page 47). The presence ot the carboxyl group explains the general resemblance to the fatty acids. Thus it forms with Potash and Soda salts which closely resemble the "soaps," and give a "lather" in aqueous solution, are "salted out" by means of solutions of common salt, and are hydrolysed by addition of excess of water.

Abietic acid appears to exist in Rosin in varying amounts, and from some specimens none was able to be separated.

§ 246. Properties.

Rosin is insoluble in water. It dissolves in alcohol [1 in 10 of a 70 per cent. strength spirit, and in its own weight of alcohol at 60° C.], ether, chloroform, glacial acetic acid and fixed oils. The percentage of unsaponifiable matter varies according to the extent of heating during distillation and consequent decomposition into hydrocarbons.

As previously stated above, Rosin forms soluble salts with alkalies resembling the "soaps."

These are soluble also in alcohol, but only very sparingly in Ether. Precipitates are produced by solutions of salts of the other metals. The zinc and copper "rosinates" are soluble in ether, while calcium "rosinate" is insoluble.

On dry distilling rosin, "rosin spirit" and "rosin oils" are obtained (v. §§ 249, 250).

As rosin is used in conjunction with oils and fats, its detection and determination in the latter is important.

It is best detected by the *Liebermann Storch* test (Practical section) and its determination is based on the fact that it forms no esters on treatment with absolute alcohol and hydrochloric acid [" *Twitchell Method*"]. For details of method, see Practical section.

[1] Levy, *Berichte*, 1909, 4305.

§ 247. Uses.

In conjunction with oils and fats in soap-making. It is employed in both soft and hard soaps, and gives a free lathering property to them. In soft soap, used to the extent of 7—10 per cent., it has a brightening and clearing effect. Cheap soft soaps are commonly heavily "rosined." It is also used as an adulterant of the higher priced varnish gums, and for stiffening waxes, and to a less extent in many other industries.

§ 248. Derivatives of Rosin.

Rosin is distilled in fire heated stills into which passes superheated steam, or by means of superheated steam alone. The steam carries the vapours resulting from the decomposition of the rosin into the condenser, and prevents the formation of an excessive amount of gas.

At about 150° C. a liquid distillate comes over. This is collected until the temperature reaches 200° C. when the receiver is changed.

The temperature is cautiously raised until it registers about 350° C. At this heat the still residue is liquid and can be drawn off. At higher temperatures coking occurs. The distillates are separated so as to yield fractions as under:

Acid water (condensed steam containing Acetic and Formic acids)	5 per cent.
Rosin spirit, crude	10 ,,
Pale yellow rosin oil	40 ,,
Blue ,, ,,	20 ,,
Green ,, ,,	25 ,,

Gases constitute about 5 per cent. and pitch about 20 per cent. of charge.

§ 249. Rosin Spirit.

The crude rosin spirit or "Pinoline" is purified by agitation with about one quarter its weight of 10 per cent. caustic soda solution. It is then redistilled.

It is a colourless mobile liquid, with a somewhat sharp, not unpleasant odour, insoluble in water. Specific gravity 0·86.

Rosin Spirit is a very complex mixture containing hydrocarbons of different boiling points with varying amounts of rosin acids and hence the spirit has no constant boiling point. The hydrocarbons so far identified belong to the Paraffins, Olefines and Benzenes together with Hydrobenzenes. The products however vary considerably according to whether the rosin has been dry distilled or steam distilled.

It is used as a solvent, and as a turpentine substitute—sometimes as a turpentine adulterant. The detection of small percentages in turpentine is a matter of difficulty.

§ **250.** ROSIN OILS.

The heavier fractions from the distillation of rosin are termed "rosin oils," and these vary in colour, consistency, and density, the earlier fractions being lightest, palest and most mobile. The specific gravity varies from ·92—·99 and is occasionally over 1·0. They all exhibit a strong *blue fluorescence*, and contain, besides hydrocarbons, phenols and varying amounts of free rosin acids (from 9—30 per cent.) the percentage varying according to the specific gravity of the fraction and the rate at which the distillation was conducted.

They are refined by treatment with Caustic Soda, which removes much of the colour and the rosin acids.

These refined oils are not readily distinguished from mineral oils except by their greater density (·95—1·0).

Rosin oils are mainly employed in the manufacture of rough greases for carts, etc. This is accomplished by treating them with milk of lime, the lime combining with the rosin acids present, a stiff emulsion resulting, from which the water does not separate.

They are also used to adulterate vegetable and animal oils, particularly linseed oil, and are best determined by the percentage of unsaponifiable matter, the examination for rosin and the determination of the specific gravity, which is high.

Steam-jacketted Mixing Pan

Applicable to the refinement and bleaching of oils and the "boiling" of Linseed Oil.

It is provided with both mechanical mixing device and air agitation. *a* is the air pump, actuated by the eccentric *b*, shaft *c* and pulley *d*, which is connected with belt driving gear. This also drives the mechanical mixer by means of bevel gear wheel *f*. The air pipe *e* ends in a blowing coil at bottom of mixer.

Plate II

Digester

For "rendering" fats and bones with high pressure steam.

The material is placed through the manhole *a*, on to a false bottom in the digester, which is then sealed up, and high pressure steam is admitted through pipe *b*, *c* is the safety valve, and *f* the blow-off pipe for steam and vapours at end of operation. These are led through a condenser, and any uncondensable gas burnt in a furnace, to avoid nuisances from the escape of foul odours. At the termination of the operation, the fat and condensed steam are run off through *g* and the exhausted material removed through manhole *a*.

Plate III

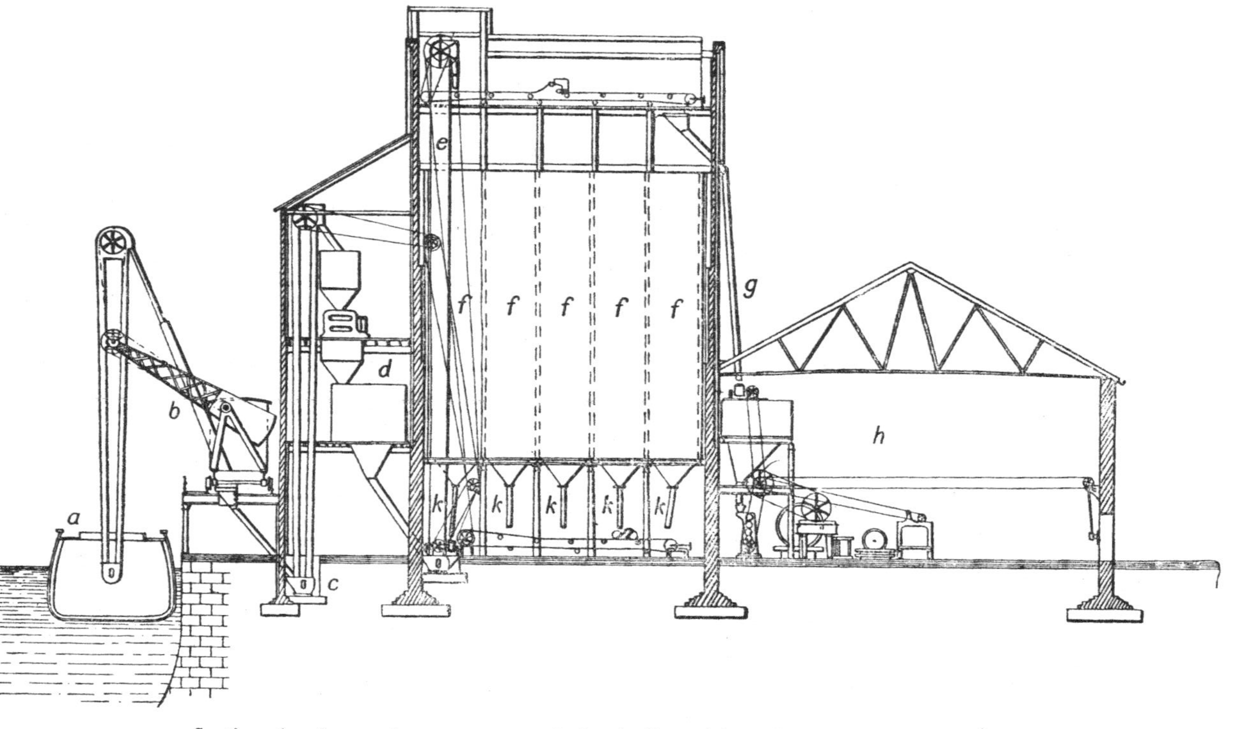

Section showing modern arrangements for dealing with seed preparatory to crushing

The seed is removed from vessel *a* by ship elevator *b*, whence it runs into *c* and is conveyed into cleaning and weighing house *d*. From thence it is elevated at *e* into the bins marked *f* in the Silo House. Each bin holds sufficient seed to be dealt with in one shift in the mill *h*. The seed discharges through openings *k* on to a travelling band and from thence is conveyed to mill through chute *g*.

Plate IV

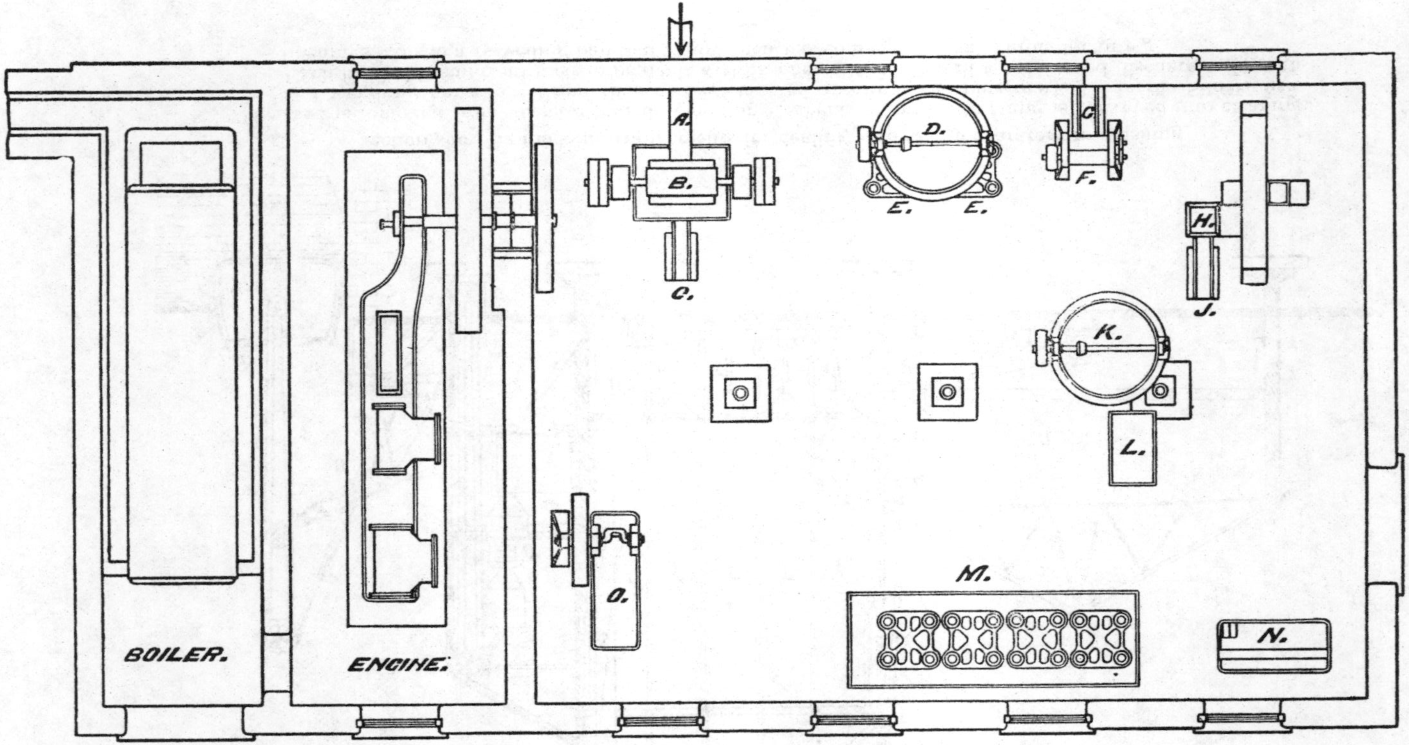

Plan of "Anglo-Premier" Mill for Oil Seeds requiring twice Pressing

The first pressing may be either hot or cold.

The first pressing is effected in the "Premier" presses *E.E.* (see Plate XXII), the second in the Anglo-American Presses *M*. The seed enters at *A* and is ground by the rolls *B*. It then passes by the elevator *C* into kettle *D*, and is pressed in *EE*. The cakes are then broken by the cake breaker *F*, and the pieces are fed into disintegrator *H*, and then to the kettle *K*, thence to moulding machine *L* and pressed at *M*. *N* is the cake-paring machine and *O* the hydraulic pumps.

a *b* *c* *d* *e* *e* *f*

"Colonial" Mill for crushing seeds requiring only once Pressing

a crushing rolls, *b* elevator, *c* kettle, *d* moulding machine, *e.e* Anglo-American type presses, *f* hydraulic pumps.

Plate VI

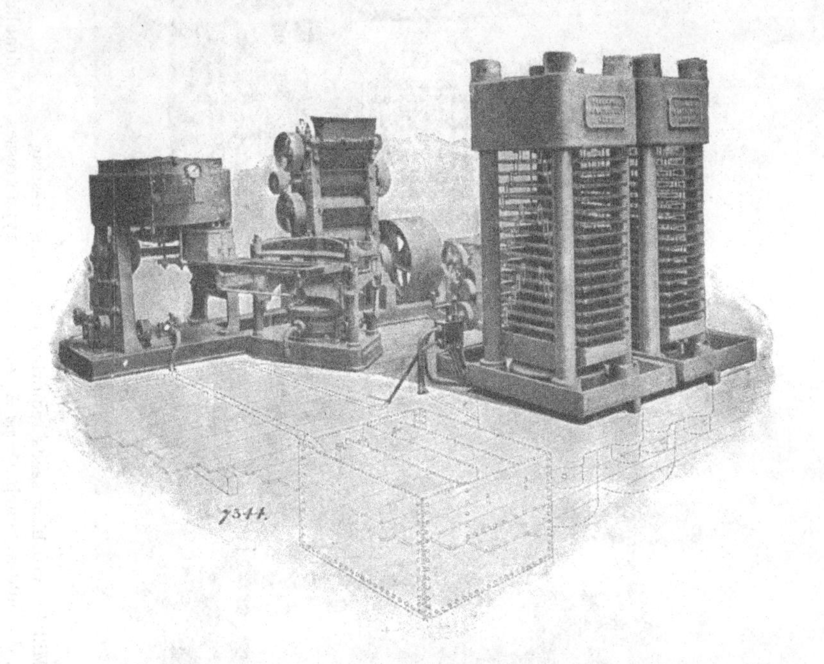

Small self-contained Anglo-American Oil Mill

Capacity 50 tons per week of 132 hours.

Applicable to Linseed, Cotton-seed, Rape, Soya beans, Mustard, Sunflower, Niger, and similar small seeds.

Underground connections shown in dotted lines.

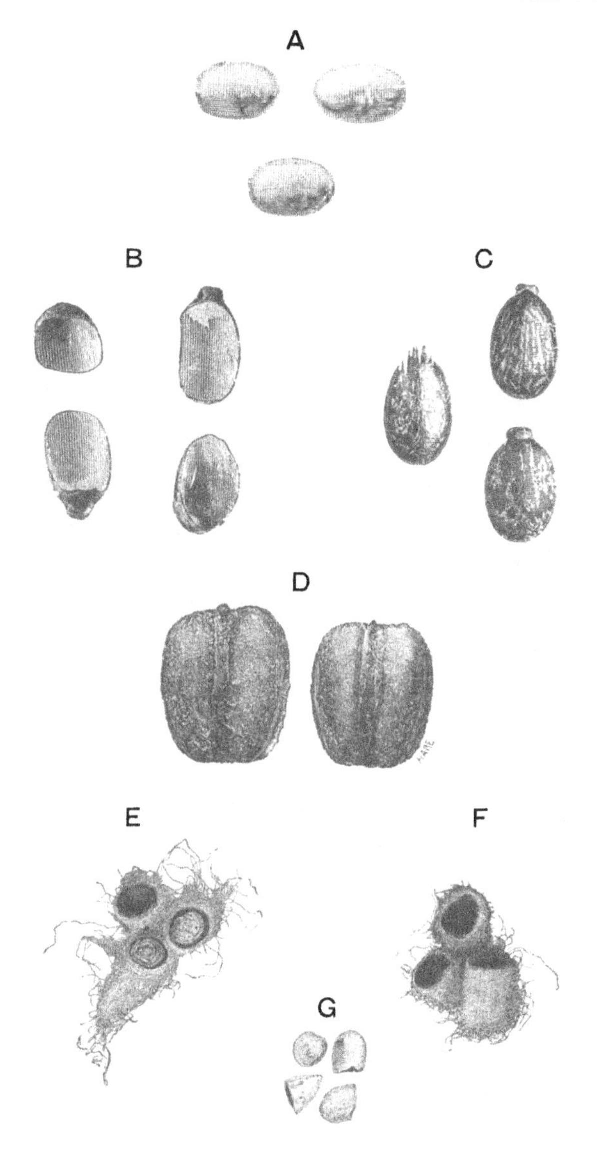

A. Decorticated *castor seeds* ready for crushing. *B*. Outer shells removed. *C*. Undecorticated castor seed. *D*. Castor Pods as received at mill. This pod or outer shell requires removal in the "Sheller." *E*. Husks of *cotton-seed* with adhering cotton. Husk and kernel cut through by machine. *F*. Empty husks. *G*. Oily kernels decorticated.

Plate VIII

Rotary Separator

For removing the husk from the oily kernel, after it has passed through the decorticator.

Used mainly for *cotton-seed.*

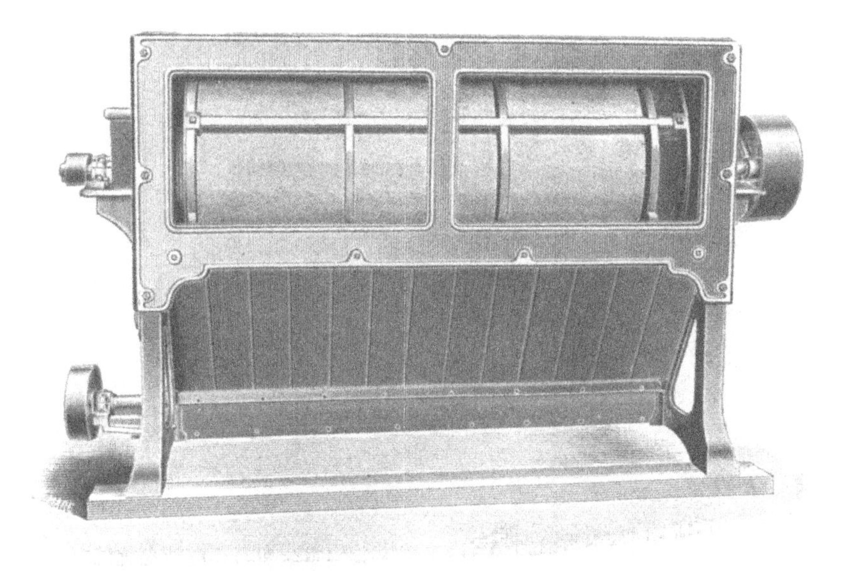

(Messrs Rose, Downs and Thompson)

Screen for Cleansing Seeds

The barrel revolves in an inclined position, the upper portion being covered with wire fine enough to retain the seed, but allowing the dust to escape. The lower end of the screen is covered with wire of a mesh which permits the seed to pass through, but retains any lumps or pieces of metal. These pass out at the end of the barrel.

Capacity 28 cwt. of seed per hour. Suitable for linseed and other small seeds.

Plate X

Cotton-seed Delinter

Capacity 6—8 tons per 24 hours.

This machine is provided with an automatic feed, and the supply is self-regulated to maintain a constant maximum flow through the machine.

(Messrs Rose, Downs and Thompson)

Castor-seed Decorticator and Separator

Capacity 1 ton of seed per hour.

Plate XII

Decorticator for Cotton-seed

The fixed member is termed the "concave." It has 31 feed knives and is held to the frame at the top by adjustable rods, and at the bottom is swung on an eccentric shaft, which provides for adjustment.

The shaft carries 14 revolving knives on a cylinder.

The chief necessity is for a uniform feed so that the seed is evenly distributed along the whole length of the knives, in order that the same amount of seed shall be treated at each revolution of the cylinder. An improved feed-hopper is provided for this purpose.

Capacity 40 tons of seed per 24 hours.

Speed of cylinder 900 revolutions per minute.

Plate XIII

Reducing Mill for Coconut and Copra

Provided with hopper feed and containing two pairs of crushing rolls adjustable to reduce to any desired degree of fineness.

Plate XIV

(Messrs Rose, Downs and Thompson)

Bone-Crushing Mill

Output 20—40 tons per day.

This mill will take any size of bones and reduce at one operation with preliminary breakage or treatment. It contains two pairs of rollers, the lower pair furnished with "toothed concaves." This enables the fineness of the product to be adjusted from half-inch pieces to "bonc dust."

Edge Runners

These are sometimes employed in connection with Anglo-American Rolls for further grinding of seeds, and for Nuts, Feeding-cake, etc.

Plate XVI

Anglo-American Seed-Crushing Rolls

The rolls are constructed of specially chilled iron pressed by hydraulic means on to the steel shafts. The seed passes into the hopper and is fed on to the rolls by a fluted feed-roller. A guide plate carries it between the first two rolls, and after being thus partially crushed, a second guide plate passes it between the second and third rolls and so on till the fifth roll is reached. It is thus crushed four times. The pressure of the rolls is maintained by a combination of springs. The grooves on the first pair of rollers are adjusted to the size of the seed treated, being cut close together for small seeds, and further apart for the larger varieties.

Plate XVII

Anglo-American Kettle and Moulding Machine

The kettle is lined on the outside with felt enclosed in iron sheeting. The stirrer is actuated by bevel wheels and shaft. The kettle is steam jacketted and is provided with a steam spray for moistening the seed. A steam pressure gauge registers the pressure on the jacket. The cooked "meats" are conveyed to the moulding machine by a sliding box filled from the bottom of the kettle. The kettle is constructed of steel plates.

Plate XVIII

Hydraulic Cake-forming and Moulding Machine

These machines take the "meats" from the kettle, and, after the press cloth is adjusted, press it into the form of a cake. This is accomplished by means of a hydraulic ram working in connection with a low pressure accumulator at about 500—700 lbs. per sq. inch. This pressure reduces the cake to about one-third of its original depth.

Output, 128 cakes per hour.

Anglo-American Hydraulic Press

In this type the plates are supported by means of racks.

Designed and tested for a pressure of 2 tons per square inch.

The columns are of mild steel.

The bottom has a large receptacle for receiving the expressed oil, from whence it is conducted to underground tanks.

Capacity 50 cwt. per 12 hours (approximately).

Plate XX

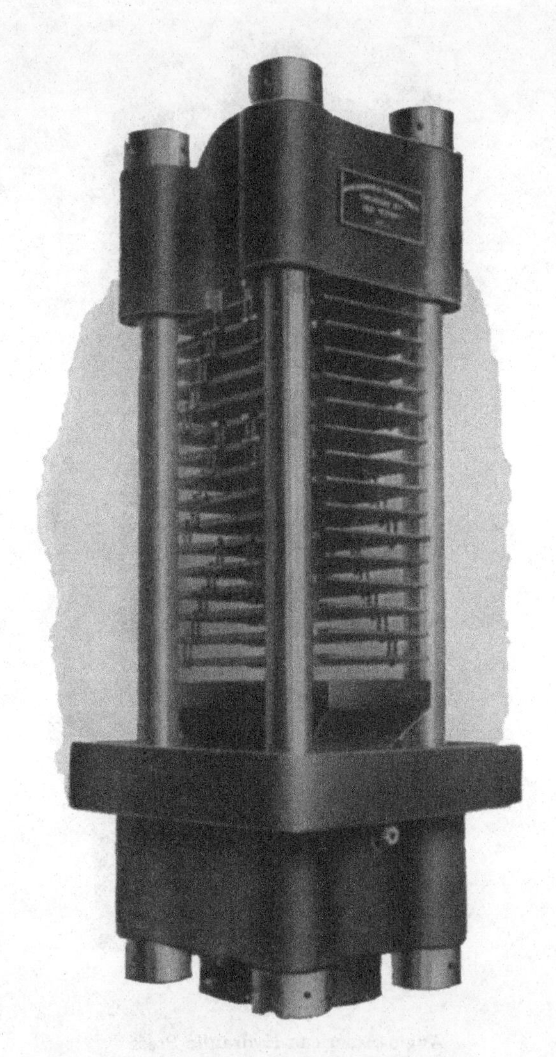

Anglo-American Hydraulic Press

In this type the plates are supported by links one from the other.

Plate XXI

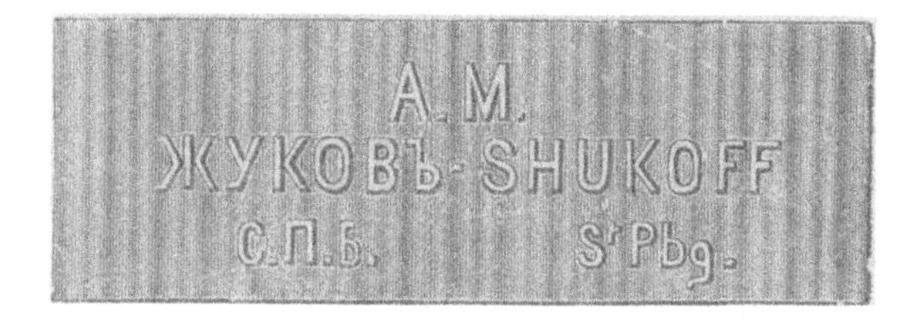

Specimens of Cake Brands used in Anglo-American Mills

The plates of the presses are designed to impress upon the finished cakes a distinctive mark or brand as shown.

Plate XXII

(*Messrs Rose, Downs and Thompson*)

"Premier" Oil Press

This type of press ("Cage" or "Clodding" Press) is used as a first-pressure press for oils such as castor, copra, palm-kernel, ground-nut.

It is combined with a kettle. The "meats" are allowed to fall into a chamber over the press which measures a constant definite quantity. A lever now closes the kettle opening and allows meal to drop into press. The ram is at its highest position, and the meal falls upon it. A disc of metal is then placed over the meal and a fresh quantity measured out, the ram meanwhile falling, until the press is full. On pressure being applied, the oil is forced out between the grooves of the cylindrical "cage."

Cage Finishing Press with removable and portable "cage." Rollers are provided for the transport of the cage from the kettle, where a preliminary pressure is given, to the finishing press.
Pressure, 350 atmospheres.

Plate XXIV

Hydraulic Pump

Provided with fast and loose pulleys and base-plate.

Suitable for working direct with hydraulic presses or through accumulators.

Driving pulleys extra large to prevent belt slip at high pressures.

The pumps are driven direct from a steel crankshaft with eccentrics turned out of the solid. Bodies constructed of mild forged steel.

Plate XXV

High, Intermediate, and Low Pressure Accumulators

These ensure uniformity of pressure in the operations, and act as safety valves for the pumps, being provided with an automatic knock-off arrangement to act immediately the accumulators reach the top of their stroke.

Plate XXVI

Set of Triple Pressure Distribution Valves

To work in connection with Accumulators (Plate XXV) for feeding 5 presses.

Plate XXVII

Anglo-American Cake Paring Machine

The *edges* of the cakes as they come from the press are rich in oil. These are therefore pared, both to increase the oil-yield and to give a neat appearance to the cake. The parings are broken up, re-ground, and placed in the kettle with the fresh charge.

Speed of knife, about 30 strokes per minute.

Plate XXVIII

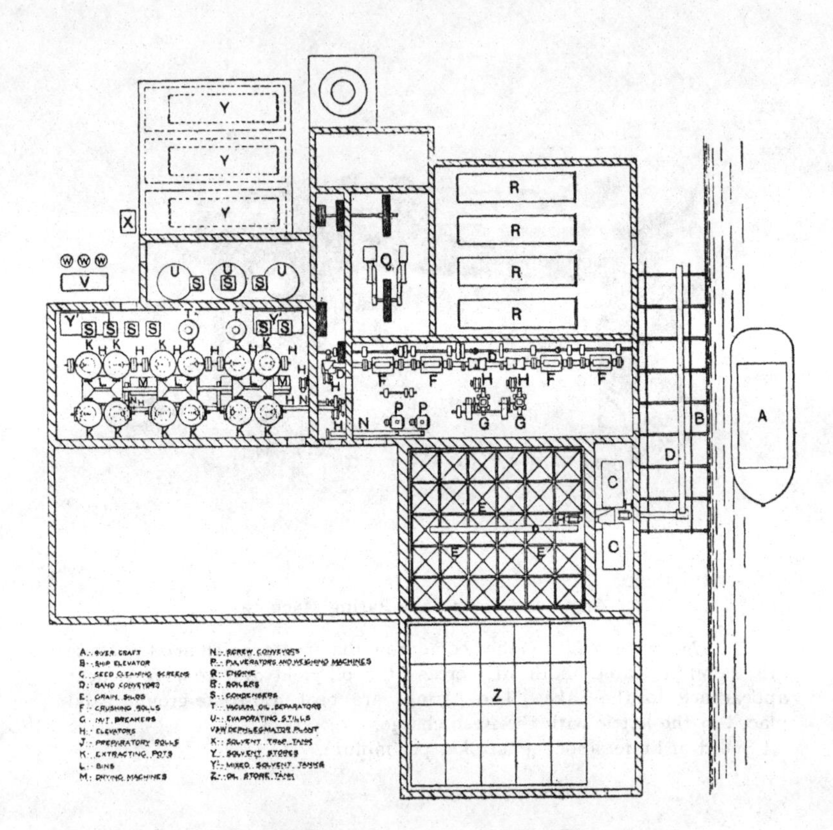

Ground Plan of Solvent-extraction Plant

Capacity 600 tons of seed per week.

Plate XXIX

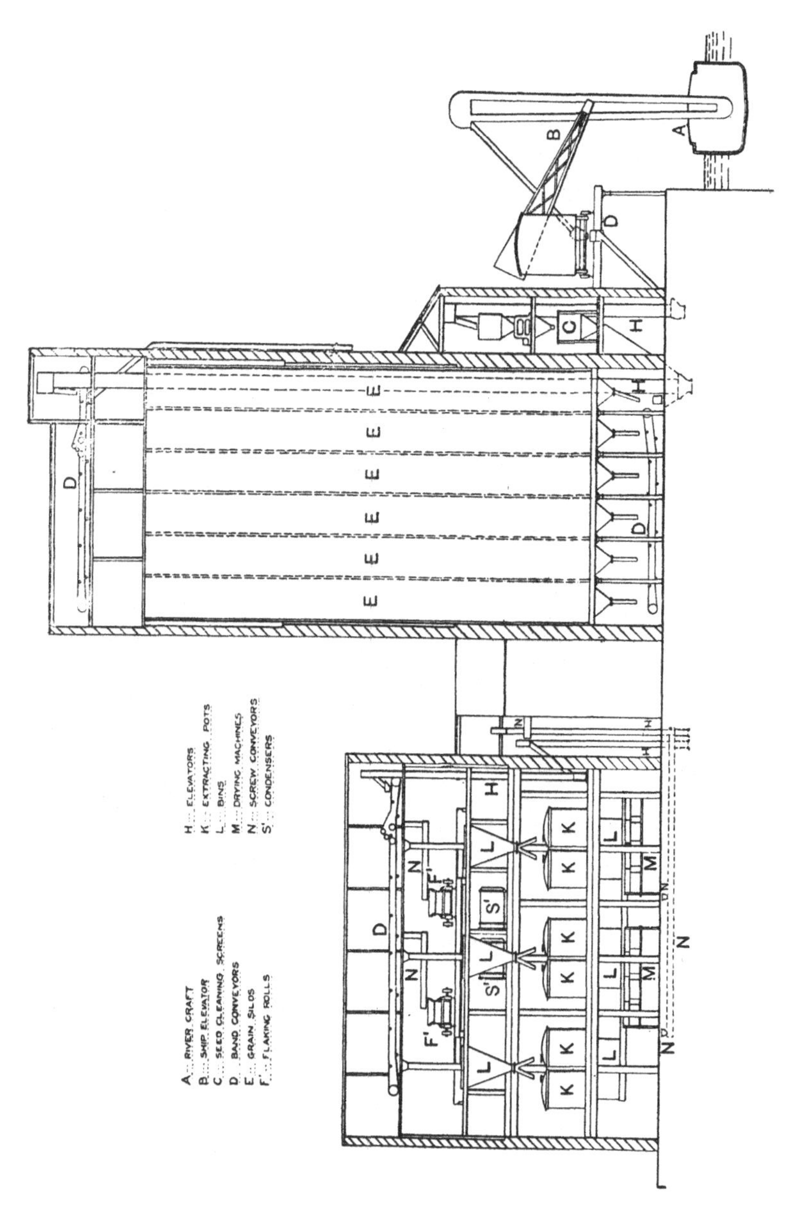

Cross Section of Solvent-extraction Plant

Capacity 600 tons of seed per week.

Plate XXX

(*Supplied by Rose, Downs and Thompson*)

View of an Extraction Plant

One of the most important points is to ensure tightness of all joints, in order to prevent waste by evaporation of solvent, and danger of fire if an inflammable solvent is used.

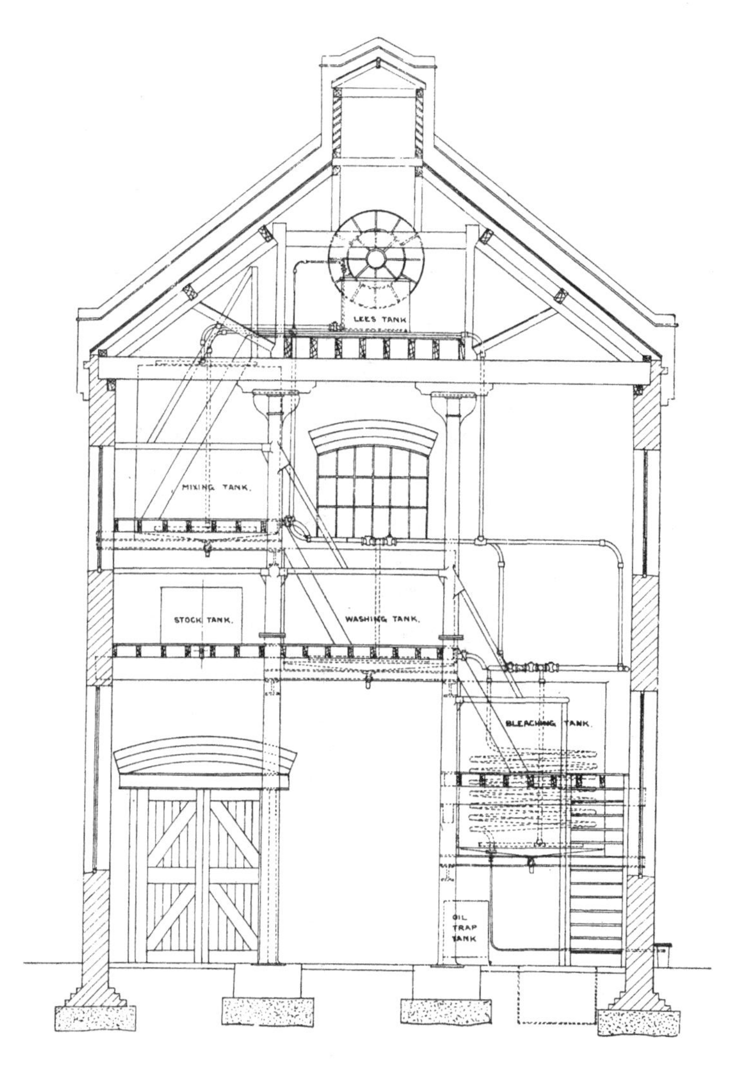

Section of Cotton-seed Oil Refinery

Plate XXXII

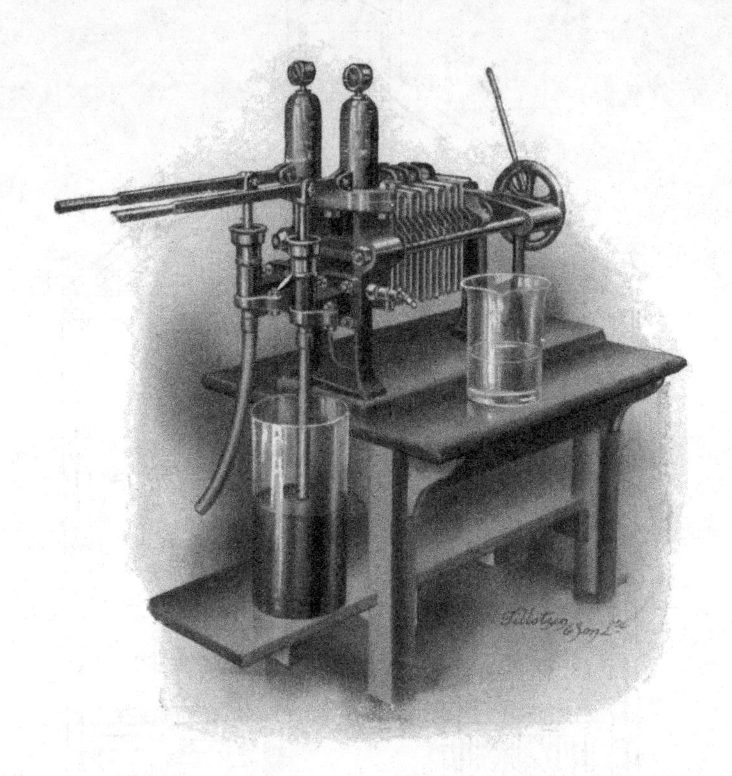

Laboratory Filter Press

For experimental trials in refining oils for edible purposes, etc. A hand-pump is employed for the filtration.

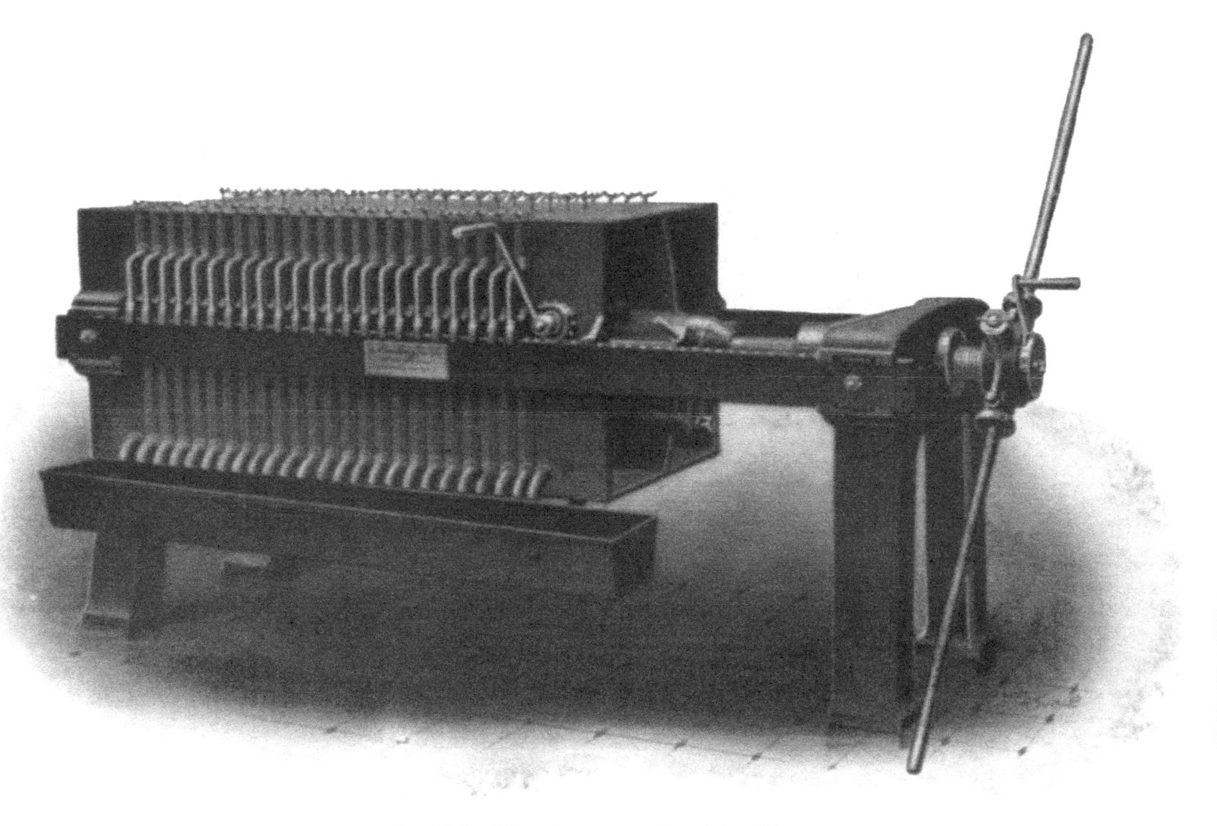

24-plate Filter Press (square-pattern) for Oils

Plate XXXIV

Filter Press for Oils, with self-contained electrically-driven pump.

(Makers, *S. H. Johnson*)

(Makers, *S. H. Johnson*)

Fullers' Earth Refining Plant for Oils

Plate XXXVI

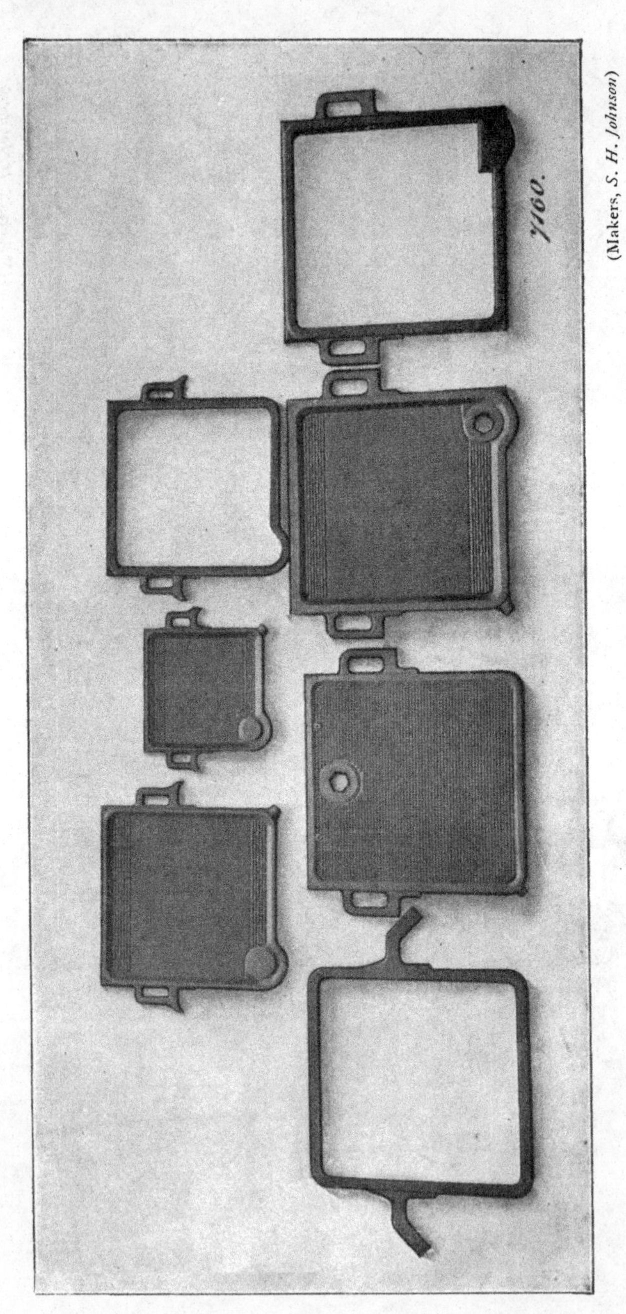

(Makers, *S. H. Johnson*)

Types of Filter Plates

Table I.

Centigrade degrees equivalent to Fahrenheit.

Fahr.	Cels.	Fahr.	Cels.	Fahr.	Cels.	Fahr.	Cels.
0	−17·8	54	+12·2	107	+41·7	160	+71·1
1	17·2	55	12·8	108	42·2	161	71·7
2	16·7	56	13·3	109	42·8	162	72·2
3	16·1	57	13·9	110	43·3	163	72·8
4	15·6	58	14·4	111	43·9	164	73·3
5	15·0	59	15·0	112	44·4	165	73·9
6	14·4	60	15·6	113	45·0	166	74·4
7	13·9	61	16·1	114	45·6	167	75·0
8	13·3	62	16·7	115	46·1	168	75·6
9	12·8	63	17·2	116	46·7	169	76·1
10	12·2	64	17·8	117	47·2	170	76·7
11	11·7	65	18·3	118	47·8	171	77·2
12	11·1	66	18·9	119	48·3	172	77·8
13	10·6	67	19·4	120	48·9	173	78·3
14	10·0	68	20·0	121	49·4	174	78·9
15	9·4	69	20·6	122	50·0	175	79·4
16	8·9	70	21·1	123	50·6	176	80·0
17	8·3	71	21·7	124	51·1	177	80·6
18	7·8	72	22·2	125	51·7	178	81·1
19	7·2	73	22·8	126	52·2	179	81·7
20	6·7	74	23·3	127	52·8	180	82·2
21	6·1	75	23·9	128	53·3	181	82·8
22	5·6	76	24·4	129	53·9	182	83·3
23	5·0	77	25·0	130	54·4	183	83·9
24	4·4	78	25·6	131	55·0	184	84·4
25	3·9	79	26·1	132	55·6	185	85·0
26	3·3	80	26·7	133	56·1	186	85·6
27	2·8	81	27·2	134	56·7	187	86·1
28	2·2	82	27·8	135	57·2	188	86·7
29	1·7	83	28·3	136	57·8	189	87·2
30	1·1	84	28·9	137	58·3	190	87·8
31	0·6	85	29·4	138	58·9	191	88·3
32	+0·0	86	30·0	139	59·4	192	88·9
33	0·6	87	30·6	140	60·0	193	89·4
34	1·1	88	31·1	141	60·6	194	90·0
35	1·7	89	31·7	142	61·1	195	90·6
36	2·2	90	32·2	143	61·7	196	91·1
37	2·8	91	32·8	144	62·2	197	91·7
38	3·3	92	33·3	145	62·8	198	92·2
39	3·9	93	33·9	146	63·3	199	92·8
40	4·4	94	34·4	147	63·9	200	93·3
41	5·0	95	35·0	148	64·4	201	93·9
42	5·6	96	35·6	149	65·0	202	94·4
43	6·1	97	36·1	150	65·6	203	95·0
44	6·7	98	36·7	151	66·1	204	95·6
45	7·2	99	37·2	152	66·7	205	96·1
46	7·8	100	37·8	153	67·2	206	96·7
47	8·3	101	38·3	154	67·8	207	97·2
48	8·9	102	38·9	155	68·3	208	97·8
49	9·4	103	39·4	156	68·9	209	98·3
50	10·0	104	40·0	157	69·4	210	98·9
51	10·6	105	40·6	158	70·0	211	99·4
52	11·1	106	41·1	159	70·6	212	100·0
53	11·7						

TABLE II.

Fahrenheit degrees equivalent to Centigrade.

Cels.	Fahr.	Cels.	Fahr.	Cels.	Fahr.
−40	−40·0	+7	+44·6	+54	+129·2
39	38·2	8	46·4	55	131·0
38	36·4	9	48·2	56	132·8
37	34·6	10	50·0	57	134·6
36	32·8	11	51·8	58	136·4
35	31·0	12	53·6	59	138·2
34	29·2	13	55·4	60	140·0
33	27·4	14	57·2	61	141·8
32	25·6	15	59·0	62	143·6
31	23·8	16	60·8	63	145·4
30	22·0	17	62·6	64	147·2
29	20·2	18	64·4	65	149·0
28	18·4	19	66·2	66	150·8
27	16·6	20	68·0	67	152·6
26	14·8	21	69·8	68	154·4
25	13·0	22	71·6	69	156·2
24	11·2	23	73·4	70	158·0
23	9·4	24	75·2	71	159·8
22	7·6	25	77·0	72	161·6
21	5·8	26	78·8	73	163·4
20	4·0	27	80·6	74	165·2
19	2·2	28	82·4	75	167·0
18	0·4	29	84·2	76	168·8
17	+1·4	30	86·0	77	170·6
16	3·2	31	87·8	78	172·4
15	5·0	32	89·6	79	174·2
14	6·8	33	91·4	80	176·0
13	8·6	34	93·2	81	177·8
12	10·4	35	95·0	82	179·6
11	12·2	36	96·8	83	181·4
10	14·0	37	98·6	84	183·2
9	15·8	38	100·4	85	185·0
8	17·6	39	102·2	86	186·8
7	19·4	40	104·0	87	188·6
6	21·2	41	105·8	88	190·4
5	23·0	42	107·6	89	192·2
4	24·8	43	109·4	90	194·0
3	26·6	44	111·2	91	195·8
2	28·4	45	113·0	92	197·6
1	30·2	46	114·8	93	199·4
0	32·0	47	116·6	94	201·2
+1	33·8	48	118·4	95	203·0
2	35·6	49	120·2	96	204·8
3	37·4	50	122·0	97	206·6
4	39·2	51	123·8	98	208·4
5	41·0	52	125·6	99	210·2
6	42·8	53	127·4	100	212·0

TABLE III.

Twaddell and Beaumé degrees and equivalent Specific Gravities.

Twaddell	Beaumé	Specific gravity	Twaddell	Beaumé	Specific gravity	Twaddell	Beaumé	Specific gravity	Twaddell	Beaumé	Specific gravity
0	0	1·000	31	19·3	1·155	66·4	36·0	1·332	102	48·7	1·510
1	0·7	1·005	32	19·8	1·160	67	36·2	1·335	103	49·0	1·515
1·4	1·0	1·007	32·4	20·0	1·162	68	36·6	1·340	104	49·4	1·520
2	1·4	1·010	33	20·3	1·165	69	37·0	1·345	105	49·7	1·525
2·8	2·0	1·014	34	20·9	1·170	70	37·4	1·350	106	50·0	1·530
3	2·1	1·015	34·2	21·0	1·171	71	37·8	1·355	107	50·3	1·535
4	2·7	1·020	35	21·4	1·175	71·4	38·0	1·357	108	50·6	1·540
4·4	3·0	1·022	36	22·0	1·180	72	38·2	1·360	109	50·9	1·545
5	3·4	1·025	37	22·5	1·185	73	38·6	1·365	109·2	51·0	1·546
5·8	4·0	1·029	38	23·0	1·190	74	39·0	1·370	110	51·2	1·550
6	4·1	1·030	39	23·5	1·195	75	39·4	1·375	111	51·5	1·555
7	4·7	1·035	40	24·0	1·200	76	39·8	1·380	112	51·8	1·560
7·4	5·0	1·037	41	24·5	1·205	76·6	40·0	1·383	112·6	52·0	1·563
8	5·4	1·040	42	25·0	1·210	77	40·1	1·385	113	52·1	1·565
9	6·0	1·045	43	25·5	1·215	78	40·5	1·390	114	52·4	1·570
10	6·7	1·050	44	26·0	1·220	79	40·8	1·395	115	52·7	1·575
10·2	7·0	1·052	45	26·4	1·225	79·4	41·0	1·397	116	53·0	1·580
11	7·4	1·055	46	26·9	1·230	80	41·2	1·400	117	53·3	1·585
12	8·0	1·060	46·2	27·0	1·231	81	41·6	1·405	118	53·6	1·590
13	8·7	1·065	47	27·4	1·235	82	42·0	1·410	119	53·9	1·595
13·4	9·0	1·067	48	27·9	1·240	83	42·3	1·415	119·4	54·0	1·597
14	9·4	1·070	48·2	28·0	1·241	84	42·7	1·420	120	54·1	1·600
15	10·0	1·075	49	28·4	1·245	84·8	43·0	1·424	121	54·4	1·605
16	10·6	1·080	50	28·8	1·250	85	43·1	1·425	122	54·7	1·610
16·6	11·0	1·083	50·4	29·0	1·252	86	43·4	1·430	123	55·0	1·615
17	11·2	1·085	51	29·3	1·255	87	43·8	1·435	124	55·2	1·620
18	11·9	1·090	52	29·7	1·260	87·6	44·0	1·438	125	55·5	1·625
18·2	12·0	1·091	52·6	30·0	1·263	88	44·1	1·440	126	55·8	1·630
19	12·4	1·095	53	30·2	1·265	89	44·4	1·445	127	56·0	1·635
20	13·0	1·100	54	30·6	1·270	90	44·8	1·450	128	56·3	1·640
21	13·6	1·105	54·8	31·0	1·274	90·6	45·0	1·453	129	56·6	1·645
21·6	14·0	1·108	55	31·1	1·275	91	45·1	1·455	130	56·9	1·650
22	14·2	1·110	56	31·5	1·280	92	45·4	1·460	130·4	57·0	1·652
23	14·9	1·115	57	32·0	1·285	93	45·8	1·465	131	57·1	1·655
23·2	15·0	1·116	58	32·4	1·290	93·6	46·0	1·468	132	57·4	1·660
24	15·4	1·120	59	32·8	1·295	94	46·1	1·470	133	57·7	1·665
25	16·0	1·125	59·4	33·0	1·297	95	46·4	1·475	134	57·9	1·670
26	16·5	1·130	60	33·3	1·300	96	46·8	1·480	134·2	58·0	1·671
26·8	17·0	1·134	61	33·7	1·305	96·6	47·0	1·483	135	58·2	1·675
27	17·1	1·135	61·6	34·0	1·308	97	47·1	1·485	136	58·4	1·680
28	17·7	1·140	62	34·2	1·310	98	47·4	1·490	137	58·7	1·685
28·4	18·0	1·142	63	34·6	1·315	99	47·8	1·495	138	58·9	1·690
29	18·3	1·145	64	35·0	1·320	99·6	48·0	1·498	138·2	59·0	1·691
30	18·8	1·150	65	35·4	1·325	100	48·1	1·500	139	59·2	1·695
30·4	19·0	1·152	66	35·8	1·330	101	48·4	1·505	140	59·5	1·700

TABLE IV.

Specific Gravities of Aqueous Solutions of Glycerin.

Glycerol per cent.	Lenz	Gerlach	
	Spec. grav. at 12°–14° C. Water at 12° C. = 1	Spec. grav. at 15° C. Water at 15° C. = 1	Spec. grav. at 20° C. Water at 20° C. = 1
100	1·2691	1·2653	1·2620
99	1·2664	1·2628	1·2594
98	1·2637	1·2602	1·2568
97	1·2610	1·2577	1·2542
96	1·2584	1·2552	1·2516
95	1·2557	1·2526	1·2490
94	1·2531	1·2501	1·2464
93	1·2504	1·2476	1·2438
92	1·2478	1·2451	1·2412
91	1·2451	1·2425	1·2386
90	1·2425	1·2400	1·2360
89	1·2398	1·2373	1·2333
88	1·2372	1·2346	1·2306
87	1·2345	1·2319	1·2279
86	1·2318	1·2292	1·2252
85	1·2292	1·2265	1·2225
84	1·2265	1·2238	1·2198
83	1·2238	1·2211	1·2171
82	1·2212	1·2184	1·2144
81	1·2185	1·2157	1·2117
80	1·2159	1·2130	1·2090
79	1·2122	1·2102	1·2063
78	1·2106	1·2074	1·2036
77	1·2079	1·2046	1·2009
76	1·2042	1·2018	1·1982
75	1·2016	1·1990	1·1955
74	1·1999	1·1962	1·1928
73	1·1973	1·1934	1·1901
72	1·1945	1·1906	1·1874
71	1·1918	1·1878	1·1847
70	1·1889	1·1850	1·1820
69	1·1858	—	—
68	1·1826	—	—
67	1·1795	—	—
66	1·1764	—	—
65	1·1733	1·1711	1·1685
64	1·1702	—	—
63	1·1671	—	—
62	1·1640	—	—
61	1·1610	—	—
60	1·1582	1·1570	1·1550
59	1·1556	—	—
58	1·1530	—	—
57	1·1505	—	—
56	1·1480	—	—
55	1·1455	1·1430	1·1415
54	1·1430	—	—
53	1·1403	—	—
52	1·1375	—	—
51	1·1348	—	—
50	1·1320	1·1290	1·1280
45	1·1183	1·1155	1·1145
40	1·1045	1·1020	1·1010
35	1·0907	1·0885	1·0875
30	1·0771	1·0750	1·0740
25	1·0635	1·0620	1·0610
20	1·0498	1·0490	1·0480
15	1·0374	—	—
10	1·0245	1·0245	1·0235
5	1·0123	—	—
0	1·0000	1·0000	1·0000

TABLE V.

Iodine Values of Unsaturated Fatty Acids and of their Glycerides.

Fatty Acid	Formula	Iodine Value of Fatty Acids	Iodine Value of		
			Monoglyceride	Diglyceride	Triglyceride
Tiglic	$C_5H_8O_2$	254·00	145·98	198·43	225·44
Hypogæic Physetoleic Lycopodic	$C_{16}H_{30}O_2$	100·00	77·44	90·07	95·25
Oleic Elaïdic Isooleic Rapic Petroselinic Cheiranthic Liver-lecithin Oleic acid	$C_{18}H_{34}O_2$	90·07	71·35	81·93	86·20
Doeglic Jecoleic	$C_{19}H_{36}O_2$	85·81	68·65	78·39	82·29
Erucic Brassidic Isoerucic	$C_{22}H_{42}O_2$	75·15	61·65	69·40	72·43
Elæomargaric Linolic Tariric	$C_{18}H_{32}O_2$	181·42	143·50	164·93	173·58
Linolenic Isolinolenic	$C_{18}H_{30}O_2$	274·10	216·47	249·02	262·15
Isanic	$C_{14}H_{20}O_2$	461·82	345·57	409·67	436·67
Therapic (?)	$C_{17}H_{26}O_2$	387·78	302·38	350·34	369·90
Clupanodonic	$C_{18}H_{28}O_2$	368·11	290·30	334·21	351·96
Ricinoleic Ricinelaidic	$C_{18}H_{34}O_3$	85·23	68·28	77·91	81·76

TABLE VI.

Saponification Values of Pure Triglycerides.

Triglyceride of Acid	Formula	Molecular Weight	Saponification Value
Acetic	$C_3H_5(O.C_2H_3O)_3$	218	772·0
Butyric	$C_3H_5(O.C_4H_7O)_3$	302	557·3
Valeric	$C_3H_5(O.C_5H_9O)_3$	344	489·2
Caproic	$C_3H_5(O.C_6H_{11}O)_3$	384	436·1
Caprylic..................	$C_3H_5(O.C_8H_{15}O)_3$	470	358·1
Capric	$C_3H_5(O.C_{10}H_{19}O)_3$	552	303·7
Lauric	$C_3H_5(O.C_{12}H_{23}O)_3$	638	263·8
Myristic..................	$C_3H_5(O.C_{14}H_{27}O)_3$	722	233·1
Palmitic..................	$C_3H_5(O.C_{16}H_{31}O)_3$	806	208·8
Daturic	$C_3H_5(O.C_{17}H_{33}O)_3$	848	198·4
Stearic	$C_3H_5(O.C_{18}H_{35}O)_3$	890	189·1
Oleic	$C_3H_5(O.C_{18}H_{33}O)_3$	884	190·4
Linolic	$C_3H_5(O.C_{18}H_{31}O)_3$	878	191·7
Linolenic	$C_3H_5(O.C_{18}H_{29}O)_3$	872	193·0
Clupanodonic	$C_3H_5(O.C_{18}H_{27}O)_3$	866	194·3
Ricinoleic	$C_3H_5(O.C_{18}H_{33}O_2)_3$	932	180·6
Arachidic	$C_3H_5(O.C_{20}H_{39}O)_3$	974	172·8
Behenic	$C_3H_5(O.C_{22}H_{43}O)_3$	1058	159·1
Erucic	$C_3H_5(O.C_{22}H_{41}O)_3$	1052	160·0
Lignoceric..............	$C_3H_5(O.C_{24}H_{47}O)_3$	1142	147·4
Cerotic	$C_3H_5(O.C_{26}H_{51}O)_3$	1226	137·3
Montanic	$C_3H_5(O.C_{28}H_{55}O)_3$	1310	128·5
Melissic	$C_3H_5(O.C_{30}H_{59}O)_3$	1394	120·7
Hydroxystearic.........	$C_3H_5(O.C_{18}H_{35}O_2)_3$	938	179·4
Dihydroxystearic	$C_3H_5(O.C_{19}H_{35}O_3)_3$	986	170·7
Trihydroxystearic......	$C_3H_5(O.C_{18}H_{35}O_4)_3$	1034	162·8
Sativic	$C_3H_5(O.C_{18}H_{35}O_5)_3$	1082	155·0
Linusic	$C_3H_5(O.C_{18}H_{35}O_7)_3$	1178	142·4

TABLE VII.

Viscosities of Glycerine Solutions.

(Archbutt and Deeley.)

Sp. Gr. at $\frac{20°}{20°}$ C.	Viscosity at 20° C. (η)	Log. of Viscosity	Differ-ences	Sp. Gr. at $\frac{20°}{20°}$ C.	Viscosity at 20° C. (η)	Log. of Viscosity	Differ-ences
1·000	·01028	$\bar{2}$·01183	—	1·047	·01283	$\bar{2}$·26078	546
1·001	·01040	$\bar{2}$·01703	520	1·048	·01846	$\bar{2}$·26625	547
1·002	·01053	$\bar{2}$·02223	520	1·049	·01870	$\bar{2}$·27173	548
1·003	·01065	$\bar{2}$·02744	521	1·050	·01893	$\bar{2}$·27722	549
1·004	·01078	$\bar{2}$·03265	521	1·051	·01917	$\bar{2}$·28272	550
1·005	·01091	$\bar{2}$·03786	521	1·052	·01942	$\bar{2}$·28823	551
1·006	·01104	$\bar{2}$·04307	521	1·053	·01967	$\bar{2}$·29375	552
1·007	·01118	$\bar{2}$·04829	522	1·054	·01992	$\bar{2}$·29928	553
1·008	·01131	$\bar{2}$·05351	522	1·055	·02018	$\bar{2}$·30482	554
1·009	·01145	$\bar{2}$·05873	522	1·056	·02044	$\bar{2}$·31037	555
1·010	·01159	$\bar{2}$·06395	522	1·057	·02070	$\bar{2}$·31593	556
1·011	·01173	$\bar{2}$·06918	523	1·058	·02097	$\bar{2}$·32150	557
1·012	·01187	$\bar{2}$·07441	523	1·059	·02124	$\bar{2}$·32708	558
1·013	·01201	$\bar{2}$·07964	523	1·060	·02151	$\bar{2}$·33267	559
1·014	·01216	$\bar{2}$·08488	524	1·061	·02179	$\bar{2}$·33827	560
1·015	·01231	$\bar{2}$·09012	524	1·062	·02207	$\bar{2}$·34389	562
1·016	·01246	$\bar{2}$·09536	524	1·063	·02236	$\bar{2}$·34953	564
1·017	·01261	$\bar{2}$·10061	525	1·064	·02266	$\bar{2}$·35519	566
1·018	·01276	$\bar{2}$·10586	525	1·065	·02296	$\bar{2}$·36087	568
1·019	·01292	$\bar{2}$·11112	526	1·066	·02326	$\bar{2}$·36657	570
1·020	·01307	$\bar{2}$·11638	526	1·067	·02357	$\bar{2}$·37229	572
1·021	·01323	$\bar{2}$·12165	527	1·068	·02388	$\bar{2}$·37803	574
1·022	·01339	$\bar{2}$·12692	527	1·069	·02420	$\bar{2}$·38379	576
1·023	·01356	$\bar{2}$·13220	528	1·070	·02452	$\bar{2}$·38957	578
1·024	·01372	$\bar{2}$·13748	528	1·071	·02485	$\bar{2}$·39537	580
1·025	·01389	$\bar{2}$·14277	529	1·072	·02519	$\bar{2}$·40119	582
1·026	·01406	$\bar{2}$·14806	529	1·073	·02553	$\bar{2}$·40703	584
1·027	·01424	$\bar{2}$·15336	530	1·074	·02588	$\bar{2}$·41289	586
1·028	·01441	$\bar{2}$·15866	530	1·075	·02623	$\bar{2}$·41877	588
1·029	·01459	$\bar{2}$·16397	531	1·076	·02659	$\bar{2}$·42467	590
1·030	·01477	$\bar{2}$·16928	531	1·077	·02695	$\bar{2}$·43059	592
1·031	·01495	$\bar{2}$·17460	532	1·078	·02732	$\bar{2}$·43653	594
1·032	·01513	$\bar{2}$·17992	532	1·079	·02770	$\bar{2}$·44249	596
1·033	·01532	$\bar{2}$·18525	533	1·080	·02809	$\bar{2}$·44847	598
1·034	·01551	$\bar{2}$·19058	533	1·081	·02848	$\bar{2}$·45447	600
1·035	·01570	$\bar{2}$·19592	534	1·082	·02887	$\bar{2}$·46049	602
1·036	·01590	$\bar{2}$·20127	535	1·083	·02928	$\bar{2}$·46653	604
1·037	·01609	$\bar{2}$·20663	536	1·084	·02969	$\bar{2}$·47259	606
1·038	·01629	$\bar{2}$·21200	537	1·085	·03011	$\bar{2}$·47867	608
1·039	·01650	$\bar{2}$·21738	538	1·086	·03053	$\bar{2}$·48478	611
1·040	01670	$\bar{2}$·22277	539	1·087	·03097	$\bar{2}$·49092	614
1·041	·01691	$\bar{2}$·22817	540	1·088	·03141	$\bar{2}$·49709	617
1·042	·01712	$\bar{2}$·23358	541	1·089	·03186	$\bar{2}$·50329	620
1·043	·01734	$\bar{2}$·23900	542	1·090	·03232	$\bar{2}$·50952	623
1·044	·01756	$\bar{2}$·24443	543	1·091	·03279	$\bar{2}$·51578	626
1·045	·01778	$\bar{2}$·24987	544	1·092	·03327	$\bar{2}$·52207	629
1·046	·01800	$\bar{2}$·25532	545	1·093	·03376	$\bar{2}$·52839	632

Viscosities of Glycerine Solutions (*continued*).

Sp. Gr. at $\frac{20°}{20°}$ C.	Viscosity at 20° C. (η)	Log. of Viscosity	Differ-ences	Sp. Gr. at $\frac{20°}{20°}$ C.	Viscosity at 20° C. (η)	Log. of Viscosity	Differ-ences
1·094	·03426	$\bar{2}$·53474	635	1·145	·08281	$\bar{2}$·91807	921
1·095	·03476	$\bar{2}$·54112	638	1·146	·08460	$\bar{2}$·92737	930
1·096	·03528	$\bar{2}$·54753	641	1·147	·08645	$\bar{2}$·93676	939
1·097	·03581	$\bar{2}$·55397	644	1·148	·08836	$\bar{2}$·94625	949
1·098	·03635	$\bar{2}$·56044	647	1·149	·09033	$\bar{2}$·95584	959
1·099	·03689	$\bar{2}$·56694	650	1·150	·09237	$\bar{2}$·96553	969
1·100	·03745	$\bar{2}$·57347	653	1·151	·09448	$\bar{2}$·97532	979
1·101	·03802	$\bar{2}$·58003	656	1·152	·09665	$\bar{2}$·98521	989
1·102	·03860	$\bar{2}$·58662	659	1·153	·09890	$\bar{2}$·99520	999
1·103	·03920	$\bar{2}$·59324	662	1·154	·1012	$\bar{1}$·00529	1009
1·104	·03980	$\bar{2}$·59989	665	1·155	·1036	$\bar{1}$·01548	1019
1·105	·04042	$\bar{2}$·60657	668	1·156	·1061	$\bar{1}$·02578	1030
1·106	·04105	$\bar{2}$·61329	672	1·157	·1087	$\bar{1}$·03619	1041
1·107	·04169	$\bar{2}$·62005	676	1·158	·1114	$\bar{1}$·04671	1052
1·108	·04235	$\bar{2}$·62685	680	1·159	·1141	$\bar{1}$·05734	1063
1·109	·04302	$\bar{2}$·63369	684	1·160	·1170	$\bar{1}$·06808	1074
1·110	·04371	$\bar{2}$·64057	688	1·161	·1199	$\bar{1}$·07893	1085
1·111	·04441	$\bar{2}$·64749	692	1·162	·1230	$\bar{1}$·08989	1096
1·112	·04513	$\bar{2}$·65445	696	1·163	·1262	$\bar{1}$·10096	1107
1·113	·04586	$\bar{2}$·66146	701	1·164	·1295	$\bar{1}$·11215	1119
1·114	·04662	$\bar{2}$·66852	706	1·165	·1329	$\bar{1}$·12346	1131
1·115	·04738	$\bar{2}$·67563	711	1·166	·1364	$\bar{1}$·13489	1143
1·116	·04817	$\bar{2}$·68279	716	1·167	·1401	$\bar{1}$·14644	1155
1·117	·04898	$\bar{2}$·69000	721	1·168	·1439	$\bar{1}$·15811	1167
1·118	·04980	$\bar{2}$·69726	726	1·169	·1479	$\bar{1}$·16990	1179
1·119	·05065	$\bar{2}$·70457	731	1·170	·1520	$\bar{1}$·18181	1191
1·120	·05152	$\bar{2}$·71193	736	1·171	·1563	$\bar{1}$·19384	1203
1·121	·05240	$\bar{2}$·71935	742	1·172	·1607	$\bar{1}$·20600	1216
1·122	·05331	$\bar{2}$·72683	748	1·173	·1653	$\bar{1}$·21829	1229
1·123	·05425	$\bar{2}$·73437	754	1·174	·1701	$\bar{1}$·23071	1242
1·124	·05520	$\bar{2}$·74197	760	1·175	·1751	$\bar{1}$·24326	1255
1·125	·05619	$\bar{2}$·74963	766	1·176	·1803	$\bar{1}$·25594	1268
1·126	·05719	$\bar{2}$·75735	772	1·177	·1857	$\bar{1}$·26875	1281
1·127	·05823	$\bar{2}$·76513	778	1·178	·1913	$\bar{1}$·28169	1294
1·128	·05929	$\bar{2}$·77298	785	1·179	·1971	$\bar{1}$·29476	1307
1·129	·06038	$\bar{2}$·78090	792	1·180	·2032	$\bar{1}$·30796	1320
1·130	·06150	$\bar{2}$·78889	799	1·181	·2096	$\bar{1}$·32130	1334
1·131	·06265	$\bar{2}$·79695	806	1·182	·2162	$\bar{1}$·33478	1348
1·132	·06384	$\bar{2}$·80508	813	1·183	·2231	$\bar{1}$·34840	1362
1·133	·06506	$\bar{2}$·81328	820	1·184	·2302	$\bar{1}$·36216	1376
1·134	·06631	$\bar{2}$·82156	828	1·185	·2377	$\bar{1}$·37606	1390
1·135	·06760	$\bar{2}$·82992	836	1·186	·2455	$\bar{1}$·39010	1404
1·136	·06892	$\bar{2}$·83836	844	1·187	·2537	$\bar{1}$·40428	1418
1·137	·07029	$\bar{2}$·84688	852	1·188	·2622	$\bar{1}$·41860	1432
1·138	·07169	$\bar{2}$·85548	860	1·189	·2711	$\bar{1}$·43307	1447
1·139	·07314	$\bar{2}$·86416	868	1·190	·2803	$\bar{1}$·44769	1462
1·140	·07463	$\bar{2}$·87292	876	1·191	·2900	$\bar{1}$·46246	1477
1·141	·07617	$\bar{2}$·88177	885	1·192	·3002	$\bar{1}$·47738	1492
1·142	·07775	$\bar{2}$·89071	894	1·193	·3108	$\bar{1}$·49245	1507
1·143	·07939	$\bar{2}$·89974	903	1·194	·3219	$\bar{1}$·50767	1522
1·144	·08107	$\bar{2}$·90886	912	1·195	·3335	$\bar{1}$·52304	1537

Viscosities of Glycerine Solutions (*continued*).

Sp. Gr. at $\frac{20^\circ}{20^\circ}$ C.	Viscosity at 20° C. (η)	Log. of Viscosity	Differ-ences
1·196	·3456	$\bar{1}$·53856	1552
1·197	·3583	$\bar{1}$·55424	1568
1·198	·3716	$\bar{1}$·57008	1584
1·199	·3856	$\bar{1}$·58608	1600
1·200	·4002	$\bar{1}$·60224	1616
1·201	·4155	$\bar{1}$·61857	1633
1·202	·4316	$\bar{1}$·63507	1650
1·203	·4485	$\bar{1}$·65174	1667
1·204	·4662	$\bar{1}$·66858	1684
1·205	·4848	$\bar{1}$·68559	1701
1·206	·5044	$\bar{1}$·70277	1718
1·207	·5250	$\bar{1}$·72013	1736
1·208	·5466	$\bar{1}$·73767	1754
1·209	·5694	$\bar{1}$·75539	1772
1·210	·5933	$\bar{1}$·77330	1791
1·211	·6186	$\bar{1}$·79140	1810
1·212	·6452	$\bar{1}$·80969	1829
1·213	·6733	$\bar{1}$·82818	1849
1·214	·7029	$\bar{1}$·84687	1869
1·215	·7341	$\bar{1}$·86577	1890
1·216	·7672	$\bar{1}$·88488	1911
1·217	·8021	$\bar{1}$·90421	1933
1·218	·8390	$\bar{1}$·92376	1955
1·219	·8781	$\bar{1}$·94354	1978
1·220	·9195	$\bar{1}$·96356	2002
1·221	·9635	$\bar{1}$·98383	2027
1·222	1·010	·00436	2053
1·223	1·060	·02516	2080
1·224	1·112	·04624	2108
1·225	1·168	·06761	2137
1·226	1·228	·08928	2167
1·227	1·292	·11126	2198
1·228	1·360	·13356	2230
1·229	1·433	·15619	2263
1·230	1·511	·17916	2297
1·231	1·594	·20248	2332
1·232	1·683	·22616	2368
1·233	1·779	·25021	2405
1·234	1·882	·27464	2443
1·235	1·993	·29946	2482
1·236	2·112	·32468	2522
1·237	2·240	·35031	2563
1·238	2·379	·37636	2605
1·239	2·528	·40284	2648
1·240	2·690	·42976	2692
1·241	2·865	·45711	2735
1·242	3·054	·48488	2777
1·243	3·259	·51306	2818
1·244	3·481	·54164	2858
1·245	3·721	·57061	2897
1·246	3·981	·59996	2935
1·247	4·263	·62969	2973
1·248	4·569	·65979	3010
1·249	4·901	·69025	3046
1·250	5·261	·72107	3082
1·251	5·653	·75224	3117
1·252	6·078	·78375	3151
1·253	6·540	·81560	3185
1·254	7·043	·84778	3218
1·255	7·591	·88029	3251
1·256	8·187	·91313	3284
1·257	8·837	·94630	3317
1·258	9·546	·97980	3350
1·259	10·32	1·01363	3383
1·260	11·16	1·04779	3416
1·261	12·09	1·08228	3449
1·262	13·10	1·11710	3482

TABLE VIII.

1918

International Atomic Weights.

	Symbol	Atomic weight		Symbol	Atomic weight
Aluminium	Al	27·1	Molybdenum	Mo	96·0
Antimony	Sb	120·2	Neodymium	Nd	144·3
Argon	A	39·88	Neon	Ne	20·2
Arsenic	As	74·96	Nickel	Ni	58·68
Barium	Ba	137·37	Niton (radium emanation)	Nt	222·4
Bismuth	Bi	208·0	Nitrogen	N	14·01
Boron	B	11·0	Osmium	Os	190·9
Bromine	Br	79·92	Oxygen	O	16·00
Cadmium	Cd	112·40	Palladium	Pd	106·7
Cæsium	Cs	132·81	Phosphorus	P	31·04
Calcium	Ca	40·07	Platinum	Pt	195·2
Carbon	C	12·005	Potassium	K	39·10
Cerium	Ce	140·25	Praseodymium	Pr	140·9
Chlorine	Cl	35·46	Radium	Ra	226·0
Chromium	Cr	52·0	Rhodium	Rh	102·9
Cobalt	Co	58·97	Rubidium	Rb	85·45
Columbium	Cb	93·5	Ruthenium	Ru	101·7
Copper	Cu	63·57	Samarium	Sa	150·4
Dysprosium	Dy	162·5	Scandium	Sc	44·1
Erbium	Er	167·7	Selenium	Se	79·2
Europium	Eu	152·0	Silicon	Si	28·3
Fluorine	F	19·0	Silver	Ag	107·88
Gadolinium	Gd	157·3	Sodium	Na	23·00
Gallium	Ga	69·9	Strontium	Sr	87·63
Germanium	Ge	72·5	Sulphur	S	32·06
Glucinum	Gl	9·1	Tantalum	Ta	181·5
Gold	Au	197·2	Tellurium	Te	127·5
Helium	He	4·00	Terbium	Tb	159·2
Holmium	Ho	163·5	Thallium	Tl	204·0
Hydrogen	H	1·008	Thorium	Th	232·4
Indium	In	114·8	Thulium	Tm	168·5
Iodine	I	126·92	Tin	Sn	118·7
Iridium	Ir	193·1	Titanium	Ti	48·1
Iron	Fe	55·84	Tungsten	W	184·0
Krypton	Kr	82·92	Uranium	U	238·2
Lanthanum	La	139·0	Vanadium	V	51·0
Lead	Pb	207·20	Xenon	Xe	130·2
Lithium	Li	6·94	Ytterbium (Neoytterbium)	Yb	173·5
Lutecium	Lu	175·0	Yttrium	Yt	88·7
Magnesium	Mg	24·32	Zinc	Zn	65·37
Manganese	Mn	54·93	Zirconium	Zr	90·6
Mercury	Hg	200·6			

INDEX

Printed in the United States
By Bookmasters